BIOTECHNOLOGY
INTELLIGENCE
UNIT

STUDIES IN VIABLE CELL IMMOBILIZATION

BIOTECHNOLOGY
INTELLIGENCE
UNIT

STUDIES IN VIABLE CELL IMMOBILIZATION

Colin Webb
George A. Dervakos
University of Manchester Institute
of Science and Technology (UMIST)
Manchester, United Kingdom

Academic Press

R.G. LANDES COMPANY
AUSTIN

Biotechnology Intelligence Unit

STUDIES IN VIABLE CELL IMMOBILIZATION

R.G. LANDES COMPANY
Austin, Texas, U.S.A.

This book is printed on acid-free paper.

Please address all inquiries to the Publisher:
R.G. Landes Company
909 Pine Street, Georgetown, Texas, U.S.A. 78626
Phone: 512/ 863 7762; FAX: 512/ 863 0081

Academic Press, Inc.
525 B Street, Suite 1900, San Diego, California, U.S.A. 92101-4495

United Kingdom Edition published by Academic Press Limited
24-28 Oval Road, London NW1 7DX, United Kingdom

International Standard Book Number (ISBN): 0-12-739225-4

Printed in the United States of America

While the authors, editors and publisher believe that drug selection and dosage and the specifications and usage of equipment and devices, as set forth in this book, are in accord with current recommendations and practice at the time of publication, they make no warranty, expressed or implied, with respect to material described in this book. In view of the ongoing research, equipment development, changes in governmental regulations and the rapid accumulation of information relating to the biomedical sciences, the reader is urged to carefully review and evaluate the information provided herein.

Library of Congress Cataloging-in-Publication Data

Webb, Colin.
     Studies in viable cell immobilization/Colin Webb, George A. Dervakos.
       p. cm. — (Biotechnology intelligence unit)
    Includes bibliographical references and index.
    ISBN 0-12-739225-4 (alk. paper)
    1. Immobilized cells. I. Dervakos, George A. II. Title. III. Series.
QH585.5.I45W43 1996
660'.6—dc20
                                       95-46476
                                         CIP

PUBLISHER'S NOTE

R.G. Landes Company publishes six book series: *Medical Intelligence Unit, Molecular Biology Intelligence Unit, Neuroscience Intelligence Unit, Tissue Engineering Intelligence Unit, Environmental Intelligence Unit* and *Biotechnology Intelligence Unit.* The authors of our books are acknowledged leaders in their fields and the topics are unique. Almost without exception, no other similar books exist on these topics.

Our goal is to publish books in important and rapidly changing areas of bioscience for sophisticated researchers and clinicians. To achieve this goal, we have accelerated our publishing program to conform to the fast pace in which information grows in bioscience. Most of our books are published within 90 to 120 days of receipt of the manuscript. We would like to thank our readers for their continuing interest and welcome any comments or suggestions they may have for future books.

Deborah Muir Molsberry
Publications Director
R.G. Landes Company

DEDICATION

To Ann, Richard and Kate Webb and Miranda, Ryan and Edmund Dervakos. We don't expect you to read it but it is for you.

CONTENTS

PREFACE

Despite decades of research and development in the field of immobilized cell systems, no book has yet been published which is dedicated solely to the use of viable cells. Furthermore, it is rare that both theoretical and practical aspects of the technology are addressed together. We hope that this book goes some way towards providing remedy for this.

The book, written essentially in two parts, begins with an evaluation of the merits of the technology, providing a comprehensive review with supporting examples taken from the authors' laboratory, where biomass support particle (BSP) systems have been studied extensively. Examples include a variety of commercially relevant large scale fermentation reactions (ethanol and beer production, production of various enzymes, organic acids and antibiotics, and environmental applications). Progress in the design of immobilized cell reactors, with particular emphasis on mixing, fluidization characteristics and carrier stability, is also covered here.

The second part of this book constitutes the latest in theoretical thinking, presenting new advances in the theoretical understanding of the mechanisms underlying the behavior of viable cells in the immobilized state. Mass transfer and reaction interactions are examined and a novel modeling approach is introduced which reflects the dynamic nature and the spatial arrangements of viable immobilized cells.

The material contained in this book has been distilled from more than fifteen years of research and development carried out in the Chemical Engineering Department at UMIST. A technology, invented by Bernard Atkinson in the late 1970s particularly for wastewater treatment applications inspired a host of subsequent investigations into a wide range of different biological systems. Some of these were referred to in a previous book *Process Engineering Aspects of Immobilized Cell Systems* (Webb C., Black G.M. and Atkinson B., Pergamon Press, 1984). The current book brings the work of the group up to date, and reflects important developments since the early days.

In this book, we have taken the rather unusual step of including "example boxes" in which we embellish the key points of the main text through the use of practical examples from our own research. These examples largely contain unpublished work from the development of BSP technology in our own laboratory and are, essentially, mini-papers in their own right. While they are intimately linked with the main text, they are also self-contained. It is our hope that the use of these examples will assist the reader in gaining better insight into BSP technology in particular and viable cell immobilization in general.

EXAMPLES

Throughout this book we have supplemented the main text with examples of research results from our own laboratory. These examples largely contain unpublished work from the development of Biomass Support Particle (BSP) technology in our own laboratory and are, essentially, mini-papers in their own right. While they are intimately linked with the main text, they are also self-contained. It is our hope that the use of these examples will assist the reader in gaining better insight into BSP technology in particular and viable cell immobilization in general.

A number of common abbreviations or acronyms appear in several of these examples as well as in the main text. The key ones are listed below.

BSP   Biomass support particle

CBR   Circulating bed reactor

FBR   Fluidized bed reactor

IPSEP   Integrated production and separation process

ppi   pores per linear inch (measure of pore size in polyurethane foam BSPs)

STR   Stirred tank reactor

ACKNOWLEDGMENTS

We have a long list of research students and assistants that we would like to thank for their contribution to the development of BSP technology in our laboratory in the department of Chemical Engineering at UMIST. At the risk of missing a few names, we would like to thank Fiona Morgan (to whose memory chapter 3 has been dedicated), Jonathan Dean, Fan-Chiang Yang, Wan-Chin Yu, Csaba Szabo, Paulo Rodrigues, Hsiang-Yu Chen, Mark Kelly, Helene Armentia, Frank Heinzelmann, Elba Bon, Anne McGrath, Don Oldfield, Amir Mansour, Elena Jimenez, Nikolaos Georgopoulos, Anders Jensen, Mehdi Nemati, Toro Nakahara, Tilsa Matthews and Hideki Fukuda. We would also like to thank Deborah Swift and Wang Ruo Hang for their assistance in the preparation of the manuscript.

CHAPTER 1

VIABLE CELL IMMOBILIZATION

Many advantages have been claimed over the years for the use of immobilized cells, both as enzyme systems and as whole viable cell systems for complete fermentation reactions. However, few of the claims have been fully substantiated, and may not even be entirely justified. Most research is involved with single applications, therefore the best that can be hoped for is some evidence that immobilized cells in each of these individual cases display some advantage over the equivalent free cell system. The purpose of this book is to consider the merits of viable cell immobilization through a collection of case studies carried out in our laboratory since the late 1970s.

The distinction between immobilized cell fermentation and immobilized cell biocatalysis is seldom made and though they are conceptually quite different. Unlike immobilized enzyme systems, immobilized viable cells can be used to carry out conventional fermentations. Microbial cells which would otherwise be freely dispersed (in almost colloidal suspension) within the fermentation environment can be encouraged to become attached in some way to a support (carrier) thus producing a discrete particulate solid phase. Such immobilization offers several potential advantages of a process engineering nature to the fermentation system. These include ease of handling and of cell separation, and lowering of bulk viscosity, as well as the obvious potential benefits of increased cell concentration.

Since the first review papers on microbial cell immobilization were publishedin the late 1970s, there has been considerable growth in research activity related to the application of immobilized viable cells to fermentation processes. Atkinson et al[1] in 1979 claimed:

"...It is expected that particles of almost any shape or size can be filled with virtually any organism, including those not generally viewed as flocculating... These are analogous to the catalyst support particles used extensively in industrial chemical reactors and the implications are that they can be used in a similar manner..."

There has, however, been relatively little interest shown by industrial fermentation technologists. Several reasons might be suggested for this reluctance on the part of industry, but probably the major cause is a misinterpretation of what is entailed by cell immobilization, in the context of fermentation technology, as distinct from enzyme immobilization. It is therefore worthwhile, in discussing the role of immobilized cells in fermentation to consider the origins of the technology.

HISTORY

During the 1950s, the realization that enzymes could be rendered water-insoluble whilst retaining their catalytic activity, led to a resurgence of interest in the possibility of using enzymes for industrial bioconversions. Thus, enzyme immobilization became a popular area of research during the 1960s, leading to the first industrial application in 1969.[2] As a result, the whole field of enzyme engineering expanded rapidly during the 1970s and applications involving multiple enzymes and co-factors were investigated. With the number of available immobilization techniques increasing markedly, and the interest in multienzyme systems growing, it was not long before the direct immobilization of whole microbial cells was attempted. By the late 1970s a number of review articles dedicated to microbial cell immobilization had been published, notably those of Abbott,[3,4] Jack and Zajic[5] and Venkatasubramanian and Vieth.[6] However, the vast majority of papers referred to in these reviews were concerned with the immobilization of cells purely for their catalytic activity and not for fermentation purposes. Cell immobilization was, then, a term used in enzyme engineering to describe the confinement or localization of microbial cells to a certain defined region of space such that they retained their catalytic activities and could be used repeatedly and continuously.[2] While enzyme technologists were discovering that more and more (in fact, all) of the cell's activity could be retained during and after immobilization, fermentation technologists were being told of the significance of microbial films in bioreactors[7] and of the advantages of cell retention in general.[8] The term Completely Mixed Microbial Film Fermenter (CMMFF) introduced by Atkinson and Davies[9] describes what might now be referred to as an immobilized cell bioreactor. It was not until the late 1970s, however, that cell immobilization techniques in general were applied to fermentation processes.[10]

Fermentation processes, by and large, involve the mass culture of micro-organisms and the products of fermentation are the products of the complex metabolic activities of the cultured cells. Thus, although there is potential, theoretically, for the replacement of fermentation processes by complex multienzyme biocatalysis, in reality, most industrially important biological products currently produced by fermentation will continue to be produced this way for some time yet. Hence, because of its predominantly enzyme engineering background, cell immobilization is seen by most fermentation technologists as something for the very distant future, if at all. It is not generally realized that cells can be immobilized very simply and exploited beneficially, for fermentation purposes, without necessitating changes to cell physiology or to overall reaction schemes. Indeed, some immobilization techniques can even be exploited within existing fermentation equipment.[11]

In fermentation, there are several benefits to be had merely from being able to treat the cells as a separate phase from the substrate/product broth. Fermentation technologists may not be aware of these or of the considerable research activity in the area simply because of the confusion with immobilized enzyme technology. Clearly there is a need to distinguish between cell immobilization for enzyme activity and cell immobilization for the purpose of carrying out full fermentation reactions. If the analogy between catalysts and enzymes is used then we might invoke the fact that the catalyst (or enzyme) is neither created nor consumed during the catalyzed reaction. While this is true for simple enzyme reactions it is generally not the case with fermentations, where the enzymes themselves are major products. Therefore, since fermentation is not biocatalysis it would, perhaps, be better to refer to cells immobilized for their enzyme activity as biocatalysts or immobilized biocatalysts and leave the term immobilized cells to describe preparations where the full metabolic activity of the cell is preserved. There would, of course, still be a gray area between immobilized cell biocatalysis and immobilized cell fermentation but this would be a genuine overlap and not just a misunderstanding of terms.

DEFINITION

When a microbial population exists in the form of single cells dispersed throughout a fermentation medium its physical behavior is governed by the properties of the bulk liquid. In other words the individual cells behave as elements of the fluid within which they are suspended. Hence, when the liquid is removed from the vessel so too are the cells. This represents a severe limitation to the operation of such systems since it is often desirable that the cells be retained for further (continuous or repeated batch) use. In order for cells to be retained they must be separated from the fermentation broth. This can be achieved most readily if the cells can be arranged so as to exhibit physical (hydrodynamic) characteristics which differ from those of the broth, in which case they can be thought of as being immobilized. Cell immobilization then can be redefined, in the context of fermentation, to describe:

> *The confinement or localization of viable microbial cells to a certain defined region of space in such a way as to exhibit hydrodynamic characteristics which differ from those of the surrounding environment.*

This is most usually achieved by significantly increasing the effective size or density of the cells by aggregation or by attachment of the cells to some support surface. Thus, flocculated cells in the form of large aggregates can be considered to be immobilized if the flocs can be separated from the bulk liquid by, for example, course screens or rapid sedimentation. Equally, cells which are entrapped within a porous matrix of sufficient size or density and cells which are attached to a solid surface are examples of cell immobilization.

The immobilization of cells can be a natural process or can be induced by chemical or physical means. Examples of natural immobilization include the formation of films in wastewater treatment systems and in the production of vinegar by the "quick" process. In both these cases recognition of the concept of immobilization was preceded by industrial exploitation of the phenomenon. Only recently, through development of techniques to effect cell immobilization artificially, has awareness of the advantages that immobilized cells can offer in the operation of fermentation systems become widespread.

IMMOBILIZATION TECHNIQUES

The methods by which cells can be immobilized are many and varied and have been reviewed extensively in several books.[12-14] Many of the techniques have been taken directly from enzyme immobilization technology. The choice of a particular application, is largely governed by the desired physiological state of the cells. For fermentation purposes, cells must be whole and viable but may be growing or non-growing. Thus the range of available techniques is restricted to those which do not adversely affect cell viability. It is also likely that techniques requiring considerable cell handling and pre-processing will be less appropriate for fermentation applications than those in which immobilization takes place within the bioreactor.

The numerous techniques for achieving immobilization can be categorized according to the physical process involved, namely: attachment, entrapment, aggregation and containment.[15] These are represented schematically in Figure 1.1. The more widely used techniques fall mainly into the first two categories.

ATTACHMENT

All forms of immobilization in which cells are in some way bound to the surface of a solid support come into the category of attachment. Although the mechanisms involved are still not fully understood, natural adhesion of cells to surfaces is a widespread phenomenon and has been the subject of many studies.[16] It provides

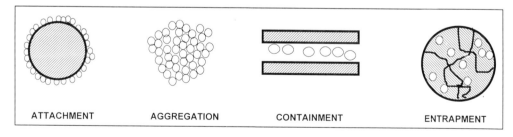

ATTACHMENT   AGGREGATION   CONTAINMENT   ENTRAPMENT

Fig.1.1. Methods for immobilizing cells.

one of the simplest techniques for cell immobilization and is the method used in both of the oldest industrial immobilized cell systems namely, vinegar production and wastewater treatment. Systems developed more recently prefer the use of particulate solid supports, often of materials such as sand or surface active porous materials such as coke or wood chips.[17] These, when used in a fluidized bed bioreactor, for example, provide a very large surface area for attachment per unit volume. The support particles, which are normally less than 1 mm in diameter, are simply placed in the bioreactor which is then inoculated in the normal way. Cells adhere to the solid surfaces and establish an active film as they grow. The film thickness may be as little as a monolayer of cells as is generally the case with animal tissue cells[18] or may be as much as several millimeters in the case of wastewater treatment organisms.[19] Cells which do not naturally adhere to surfaces can sometimes be encouraged to attach by chemical means such as cross linking by glutaraldehyde, silanization to silica supports or by chelation to metal oxides.[15] In such cases strength of attachment is similar to that in natural adhesion.

Attached cells (Fig. 1.2) are in direct contact with the surrounding environment and hence subject to any forces of shear or attrition which may result from the relative motion of particles and fluid. It is therefore likely that some cells will become detached and enter the bulk fluid phase. It is also difficult to control

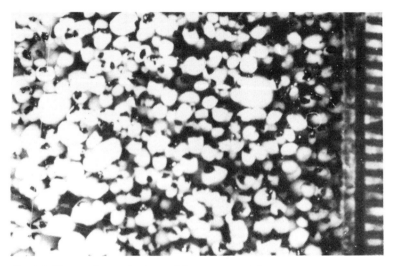

Fig. 1.2. Yeast cells attached to solid support particles.

or even determine the depth of the attached biofilm. Nevertheless, the technique is cheap and simple and the art is to provide the right surface in a suitable form for colonization by the desired organism or population of organisms in as high a density as possible.

ENTRAPMENT

Entrapment of cells can take place within a variety of porous structures which are either preformed or formed in situ around the cells. In the case of preformed structures the entrapment usually occurs as a natural consequence of cell growth and so, like natural attachment, the effectiveness of the immobilization varies with cell type and support type. Porous structures which are formed in situ, on the other hand, can be used to immobilize almost any type of cell, though the conditions under which the support particle is formed may be harmful to the cells in some cases.

Entrapment within preformed porous supports can be carried out on a microscopic level with micro-porous particles such as brick, ceramics, sintered glass or kieselguhr (pore entrapment), or on a macroscopic level with particles having relatively large pores. An example of macro-pore particles are those developed by Atkinson et al.[1] These Biomass Support Particles (BSPs) are large open structures fabricated from either stainless steel wire or reticulated foams and can be used with a wide variety of cell types (see Example 1: Biomass Support Particles). Unlike attached cells, cells which have been entrapped in preformed porous supports are protected from the shear field outside the particles but, like attached cells, they are not confined to the particle by any barrier. It is, therefore, unlikely that the bioreactor broth in such systems would remain cell-free. However, an advantage of this technique is that the immobilized biomass holdup can be controlled such that cells growing beyond the boundaries of the particle are removed by abrasion, either by the flow field surrounding the particle or by particle-particle contact.

The most popular form of cell immobilization currently in use involves the entrapment of cells within porous structures which are formed in situ around the cells (Fig. 1.3). The cells, in the form of a slurry or paste, are generally mixed with a compound which is then gelled to form a porous matrix under conditions sufficiently mild so as not to affect the viability of the cells.[20] The

majority of techniques involving in situ entrapment for fermentation purposes make use of polysaccharide gels.[21] Of these, which
include κ-carrageenan, agar and alginates, calcium alginate gel is
the most popular.

Gel entrapment provides a controlled means of achieving what
is quite a common occurrence in nature for certain organisms, e.g.
slime forming bacteria. The physical properties of gels are not dis-
similar to such slimes but whereas only a few species will form
slimes, almost any organism can be immobilized by gel-entrap-
ment. Cell growth within the particles can occur, though if the
cell concentration exceeds approximately 30% v/v the gel will lose
its integrity.[22] Calcium alginate gels are unstable in the presence
of calcium chelators, such as phosphate, and gas evolution within
the gels can also be a cause of particle disruption.[23]

Fig. 1.3. Yeast cells entrapped in Ca-alginate beads.

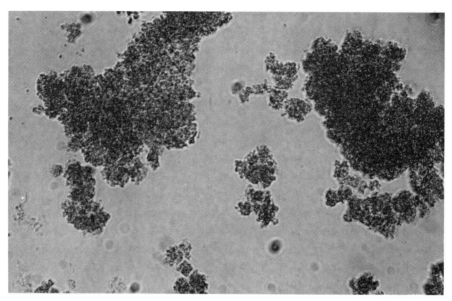

Fig. 1.4. Cells aggregated in activated sludge.

AGGREGATION

By flocculating to form large aggregates, cells may become immobilized in the sense that it then becomes possible to retain them in continuously operated bioreactors, for example, in the form of packed or fluidized beds. Natural flocculation of yeast cells is well known and has been exploited in beer production using tower bioreactors.[24] Fungal mycelia also form aggregates in the shape of spherical pellets[25] and flocculation is a key feature of the activated sludge process for wastewater treatment (Fig. 1.4).[26] Microbial species not normally regarded as flocculent can be induced to flocculate under certain conditions.[27]

CONTAINMENT

Cells may also be immobilized by containment behind a barrier, either preformed or formed in situ. The barrier may be as simple as an interface between two immiscible liquids[28] in which case the cell suspension is emulsified with an organic solvent and resuspended as droplets in an aqueous phase or may involve semipermeable membranes used for microfiltration and ultrafiltration.[29] Such systems maintain completely cell-free process liquors, while allowing nutrients to diffuse to the cells, and therefore find use in the area of mammalian tissue cell culture.[30]

METABOLIC CHANGES IN IMMOBILIZED CELLS

By immobilizing cells it becomes possible, at least in principle, for biological particles of any size, shape and density to be produced for a wide variety of both prokaryotes and eukaryotes.[1] As a consequence, one of the major features of most immobilization processes is the very high cell concentrations that can be achieved and this, combined with the ability to handle immobilized cells distinguishably from the fluid phase, offers a number of potentially advantageous possibilities for fermentation processes. In addition, a variety of claims have been made regarding the effects of immobilization on the metabolism of the cell. It appears that in certain systems immobilization may lead to the enhancement of some metabolic activity of the cell, though the scarcity of reports describing decreased metabolic rates may simply reflect the difficulty in ruling out diffusional limitations as the cause of the decrease. On the other hand an apparent enhancement of a metabolic rate may be the result of cell growth in the immobilization matrix or of advantageous diffusion limitations, e.g. substrate inhibition.

In general, there appears to be no consistent method for predicting the direction and the magnitude of the metabolic changes. For example, adsorption of yeast to various solid surfaces has been reported to affect the intrinsic growth rate of the cells which either increased[31,32] or decreased.[33] Adsorption of *Saccharomyces carlsbergensis* onto porous glass beads increased the yield of ethanol on glucose and decreased the carbon dioxide yield,[34] however, covalently bound cells had a reduced activity.[35]

Various immobilized cell systems appear to be affected by some sort of metabolic change. In the absence of specific precursors to capsaicin, immobilized cells of *Capsicum frutescens* produced between two and three orders of magnitude higher yields than did suspended cells over five-day or ten-day culture periods.[36,37] The solasodine content in immobilized cells of *Solanum xanthocarpus* was reported to increase by two-fold in comparison to free cells of the same culture stage.[38] The enhanced ajmalicine productivity of *Catharanthus roseus* in Ca-alginate were exploited in a systematic manner by simultaneously applying in situ adsorption of ajmalicine and by using elicitors.[39] The mean specific rate of hydrogen production by immobilized *Rhodospirillum rubrum* was twice the rate exhibited by free cells[40] while the specific hydrogen production by an immobilized cyanobacterium was 25% higher than free cells.[41] Callegari et al[42] reported that light was utilized more efficiently by immobilized *Chlorella sorokiniana* for the photoproduction of oxygen than by free cells. *E. coli* cells immobilized in Ca-alginate produced elevated quantities of byproducts but consumed glycerol at twice the rate of suspended cells.[43] Cells grown immobilized in alginate and then released from the gel synthesized more β-galactosidase per cell in response to induction than suspended cells. Under both aerobic and anaerobic conditions, the cell yield from glycerol for immobilized cells was half that for suspended cells. The effects of the microenvironment on yeast metabolism were studied by Chen and Wu.[44] Yields of various metabolic products and the utilization ratios of various specific amino acids were reported to differ from those of a freely suspended culture. Evans and Wang[45] reported that the production of pigments by immobilized *Monascus* species utilizing polymeric resin adsorption was higher than by free cells and this was attributed to the provision of a support for the mycelium which mimics the conditions of a solid-culture.

A number of hypotheses have been proposed in order to explain the altered metabolic behavior of cells in the immobilized state:

- Disturbances in the growth pattern
- Surface tension/osmotic pressure effects
- Reduced water activity
- Cell-to-cell communication
- Changes in the cell morphology
- Altered membrane permeability
- Media supplementation
- Diffusion limitations

DISTURBANCES IN THE CELL GROWTH PATTERN

Disturbances in the growth pattern of immobilized cells due to contact with the immobilization carrier or other cells were proposed by Doran and Bailey[33] to explain various metabolic changes of yeast cells immobilized on crosslinked gelatin in an apparently gradientless reactor. These metabolic changes included reduced biomass yields, decreased specific growth rates and increased rates of glucose consumption and ethanol production. The observed changes were attributed to disturbances to the yeast cell cycle by cell attachment, which may have caused alterations in the normal pattern of yeast bud development, DNA replication, and synthesis of cell wall components. The effect of hydroxyurea—an inhibitor of DNA synthesis—on the fermentation rates and the cell cycle operation were different in immobilized cells suggesting a disturbance of the yeast cell cycle due to immobilization.[46] Studies on the unsteady state behavior of immobilized and suspended yeast confirmed that immobilization affects cellular regulation and control.[47] ^{31}P-NMR spectroscopy of fermenting immobilized and suspended cells showed differences in intermediate metabolite levels. The internal pH of the immobilized cells was lower than the suspended cell internal pH.[48,49]

SURFACE TENSION/OSMOTIC PRESSURE EFFECTS

Changes in the surface tension and osmotic pressure ('skin effect') were suggested by Vijayalakshmi et al[50] to explain a threefold increase in the oxygen uptake rate of yeast cells when immobilized onto a crosslinked pectate via a transition metal link.

REDUCED WATER ACTIVITY

Reduced water activity and/or oxygen deficiency was proposed by Mattiasson and Hahn-Haegerdal[51] to explain changed yields and/or new metabolic behavior of immobilized cells. Polyethylene glycol was used to study the effects of decreased water activity on cell metabolism.[52] In batch experiments with *Saccharomyces cerevisiae*, a marked increase in ethanol production was observed in the initial phase of the fermentation in batches containing 12-22% polyethylene glycol (water activity = 0.99). In other experiments, the glycerol produced by *Dunaliella parva* was increased in media in which the water activity was decreased by polyethylene glycol or NaCl.

CELL-TO-CELL COMMUNICATION

Cell-to-cell communication in immobilized eukaryotic cells may be the reason for the enhancement of certain biosynthetic activities.[53,54] Haldimann and Brodelius[55] reported that alginate entrapped *Coffea arabica* cells enhanced the production of methylxanthine alkaloids in comparison to freely suspended cells under the same conditions. It was proposed that a reversible interaction between alginate and pectic acid, which is a component of the cell wall, may be established by calcium ions. The alginate could function as a 'glue' between cells and thereby mediate a cell-to-cell interaction similar to that of a differentiated plant tissue. Ayabe et al[56] reported that the induction of echinatin in *Glycyrrhiza echinata* cells could be achieved by either immobilization in alginate or by transfer of free cells into a medium containing $CaCl_2$ and addition of Na-alginate into the suspension culture.

CHANGES IN CELL MORPHOLOGY

Changes in cell morphology caused by immobilization may be responsible for enhancement in the metabolic activities of the cells. Chlortetracycline production by immobilized *Streptomyces aureofaciens* was correlated with the morphological development of cells in the gel.[57] It was observed that micropellets were formed within alginate beads which were never found in free cell culture of the mycelia. These micropellets may have enlarged the active biological surface and, hence, may be the cause of the higher productive capacity of immobilized cells. The morphological differentiation of immobilized *Claviceps paspali* mycelium during semi-continuous cultivation—which never appeared during fermentation of free mycelium—was also connected with the improved, prolonged vitality and metabolic activity of the immobilized cells.[58] Immobilized growing hybridoma cells exhibited a higher rate of oxygen uptake compared with the free cells and this was shown to be closely related with the formation of cell colonies in the gel particles.[59]

ALTERED MEMBRANE PERMEABILITY

Changes in the cell membrane permeability caused by interactions between the cell and the immobilization carrier may allow the passage of enzyme substrates which cannot enter the normal cell and/or enhance product excretion. Membrane permeability has been studied extensively in the context of plant cell immobilization. Earlier reports were concerned with the retention of the enzymatic activities of immobilized permeabilized cells.[60] Growth of intermittently permeabilized *Catharanthus roseus* cells was observed by Brodelius and Nilsson.[61] The use of chitosan/alginate and κ-carrageenan/chitosan copolymers was reported to allow the concurrent immobilization and permeabilization of plant cells while maintaining some cell viability.[62-65] Immobilized cells of *Chlamydomonas reinhardtii* photoproducing ammonium showed a significant increase in the nitrite uptake rate;[66,67] this was ascribed to a change in membrane permeability as a consequence of cell-matrix interactions.

Inclusion of sand within Ca-alginate beads was reported to enhance ethanol production and this was ascribed to the effect of Si^{4+} on the permeability of the yeast cells.[68] A noticeable difference between free and immobilized cells of *Gibberella fujikuroi* was that in the immobilized cells, all of the bikaverin was excreted into the medium, in contrast to free cells. This difference was also attributed to increased membrane permeability.

MEDIA SUPPLEMENTATION

Compounds present in an immobilization matrix may contribute to a more favorable medium composition. For example, the maximum concentrations of ethanol produced during the fermentation of 320 g/L glucose by *Saccharomyces bayanus* was higher when the yeast cells were immobilized either by adsorption on Celite or by entrapment in κ-carrageenan beads (from 10.5% with free cells to 14.5% and 13.1%, respectively). This increase was due to medium supplementation with the compounds present in the immobilization supports.[69] The degree of the support hydrophilicity seems also to play an important role in the systems' ability to reach high ethanol concentrations.[70]

DIFFUSION LIMITATIONS

Under conditions of intra-particle mass transfer limitation, cells immobilized at certain locations within the aggregate may be exposed to concentrations of substrates and products which promote a particular pathway for the flow of mass within the cell and, thus, the production of metabolites associated with this pathway. It is also possible that such conditions of concentration are not achievable in freely suspended cells due to the stoichiometrical or operational constraints imposed by the reactor type and its mode of operation.

Webb et al[71] reported a more than three-fold increase in the specific cellulase productivity of *Trichoderma viride* immobilized in stainless steel biomass support particles in comparison with the freely suspended cells; the increase was ascribed to advantageous diffusional limitations. Kopp and Rehm[72] reported that with increasing Ca-alginate concentrations, alkaloid production even higher than that of free mycelia could be observed. This was ascribed to the reduced oxygen into the beads of high Ca alginate concentrations which induced a shift in the clavine alkaloid production; 8% Ca alginate-immobilized cells produced agroclavine as the main product, whereas free cells oxidized agroclavine with oxygen to produce elymoclavine and excreted this alkaloid as the main product. Similarly Murdin et al[73] proposed that advantageous diffusion limitations may explain the higher specific antibody production in a packed-bed-reactor compared to suspension cultures.

It should be noted that claims for advantageous diffusional effects may be difficult to substantiate due to the difficulty in predicting accurately the effectiveness factor of particles containing viable cells; in certain cases the effective diffusivity may be known only within an order of magnitude, while the exact spatial profile of the biomass concentration and/or the distribution and size of microcolonies within the aggregate may be difficult to measure. Convective transfer may complicate the situation even further.

Experimental techniques are now available for use with immobilized cells however, which assist in the characterization of the metabolic state of the immobilized cells and which may indicate shifts in metabolism by monitoring various intracellular metabolites. These techniques have been reviewed by Karel et al.[74] They include radioisotope labeling with ^{35}S for the determination of overall rates of cell mass synthesis and degradation in immobilized cell reactors and ^{31}P-NMR spectroscopy for the measurement of region-specific pH and of the relative amounts of phosphorylated intermediates and products.[75,76] For example, a comparison of freely suspended and immobilized *Catharanthus roseus* cells has shown that the biosynthetic capacity of the cells does not change upon immobilization.[76] Fluorescence monitoring was used in order to both examine metabolic changes in the cells and to guard against toxins which cause visible changes in the fluorescence of immobilized cells.[77]

THE MERITS OF VIABLE CELL IMMOBILIZATION

The immobilization of viable microbial cells for fermentation purposes has been practiced, albeit unwittingly, within the fermentation industry since the introduction of the 'quick' process for vinegar production by Scheutzenbach in 1823. Recognition of the benefits of this particular process and of those of high rate biological filters used in wastewater treatment led, during the mid 1970s, to the intentional immobilization of a wide range of other microbial species for a variety of fermentation applications. The merits of using immobilized cells instead of freely suspended cells depend on the characteristics of the particular system.[78] Some advantages of immobilized cells over free cells are:

- Improved biological stability
- Improved biomass holdup
- Improved mass transfer
- Improved product yields
- Improved reactor choice
- Improved downstream processing
- Advantageous partition effects
- Improved product stability
- Advantages due to cell proximity
- Improved reaction selectivity

These merits will be explored more fully in subsequent chapters using examples from our own work.

EXAMPLE 1:
BIOMASS SUPPORT PARTICLES

The majority of techniques reported for cell immobilization may be described as being 'active' (i.e. generally requiring the production of cells which, after mixing with some chemical agent, are immobilized by chemical or physical means). 'Passive' immobilization techniques are also available, i.e. those in which films or flocs of cells form naturally around or within support material provided for that purpose. The technology described here relies on such a passive immobilization procedure, specifically the immobilization of cells within porous structures termed Biomass Support Particles (BSPs). The concept is one of providing a structure within which the organisms may grow, protected from high external shear, and within which even weakly adhesive or flocculent organisms will be retained. This technology has been applied to a variety of microbial and plant cell systems. BSPs consist of a network of support material, comprised of a complex pattern of contiguous voids which provides a high internal porosity. Two principle types are currently in use:

STAINLESS STEEL SPHERES

These are fabricated by cutting a length of stainless steel stocking and die pressing this into a sphere (Example Fig. 1.1a, 1.1b, 1.1c). Stainless steel BSPs generally have the following characteristics.

Shape: Spherical
Size: Typically 6-10 mm diameter
Voidage: 0.8
Matrix Density: 7,700 kg/m^3
Bulk Density*: 2,380 kg/m^3

*Particle density when full of biomass

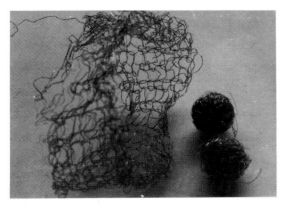

1.1a

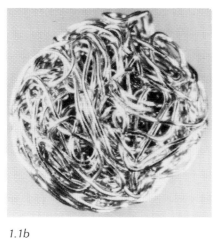

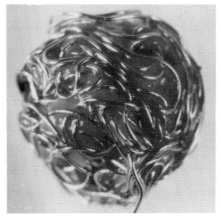

1.1b   *1.1c*

Example Fig. 1.1. Stainless steel BSPs.

RETICULATED FOAMS

These may be made from a range of materials, most commonly polyurethane, and may be cut into a variety of shapes. Reticulated foams differ from ordinary foams in that the windows between the cells have been removed either chemically or by an explosive process ('zapping'). The resultant foam has a very open structure with a high degree of interconnection between the pores (Example Fig. 1.2a, 1.2b, 1.2c). Foam BSPs are available with a variety of pore sizes, ranging usually from 10 pores per inch (ppi) to 80 ppi. Polyurethane foam BSPs generally have the following characteristics.

Shape:	Cubes, cuboids or sheets
Size:	Typically 6 x 6 x 6 mm to 25 x 25 x 10 mm
Voidage:	0.97
Matrix Density:	1,100 kg/m^3
Bulk Density*:	1,050 kg/m^3

* Particle density when full of biomass

1.2a

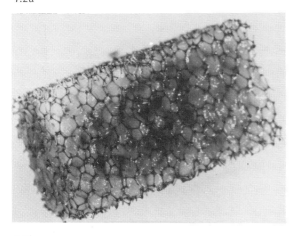

1.2b

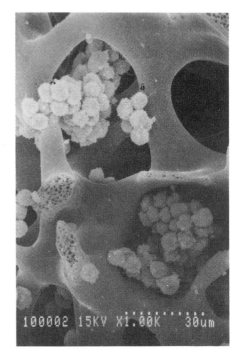

1.2c

Example Fig. 1.2. Polyurethane foam BSPs.

APPLICATIONS USING BSPs

Amongst the many applications investigated with BSPs, those listed in Example Table 1.1 have been carried out in our laboratory.

Biomass Support Particles provide a sound basis for a robust immobilized cell technology. Development is most advanced in the field of wastewater treatment, but the technology is equally advantageous in more demanding areas such as plant cell culture. Use of BSPs involves a very simple process. Typically, prior to fermenter operation, BSPs and growth medium are placed into the bioreactor, which is sterilized before being inoculated. Cells are then allowed to grow and fill the BSPs. Continuous operation is normally started after a period of batch growth.

Some advantages of BSP technology over other methods of immobilization are:
- no chemical additions are required
- no need for preproduction of cells
- no aseptic handling of particles required
- particles are reusable
- low costs, compared to some active techniques

Example Table 1.1. Applications of BSPs studied in the authors' laboratory

Organism	Application
Mixed	Wastewater Treatment (Captor® Process)
Saccharomyces sp.	Yeast growth/ethanol
Acetobacter	Acetic acid (vinegar)
Trichoderma viride	Cellulase
Saccharomyces uvarum	Beer
Aspergillus awamori	Glucoamylase
Streptomyces sp.	Antibiotics
Penicillium chrysogenum	Penicillin
Aspergillus niger	Citric acid
Xanthanmonas campestris	Xanthan gum
Lactobacillus plantarum	Lactic acid
Thiobacillus ferrooxidans	Ferrous sulphate oxidation (H_2S removal)
Mouse myeloma	Tissue culture
Various	Plant cell culture

REFERENCES

1. Atkinson B, Black GM, Lewis PJS and Pinches A. Biological particles of given size, shape and density for use in biological reactors. Biotechnol Bioeng 1979; 21:193-200.

2. Chibata I and Tosa T. Immobilized cells. Historical background. Appl Biochem Bioeng 1983; 4:1-9.

3. Abbott BJ. Immobilized cells. Ann Rep Fermentation Processes 1977; 1:205-233.

4. Abbott BJ. Immobilized cells. Ann Rep Fermentation Processes 1978; 2:91-123.

5. Jack TR and Zajic JE. The immobilization of whole cells. Adv Biochem Eng 1977; 5:125-145.

6. Venkatasubramanian K and Vieth WR. Immobilized microbial cells. Progr Ind Microbiol 1979; 15:61-86.

7. Atkinson B and Fowler HW. The Significance of microbial film in fermenters. Adv Biochem Eng 1974; 3:221-277.

8. Atkinson B. Biochemical Reactors. London. Pion Press, 1974.

9. Atkinson B and Davies IJ. The completely mixed microbial film bioreactor—a method of overcoming washout in continuous fermentation. Trans Instn Chem Engrs 1972; 50:208-216.

10. Kennedy JF and Cabral MS. Immobilized living cells and their applications. Appl Biochem Bioeng 1983; 4:189-280.

11. Gbewonyo K, Meier J and Wang DIC. Immobilization of mycelial cells on celite immobilized enzymes and cells: part B. Methods in Enzymology 1987; 135:318-333.

12. Rosevear A, Kennedy JF and Cabral MS. Immobilized Enzymes and Cells. Bristol. Adam Hilger, 1987.

13. Tampion J and Tampion MD. Immobilized Cells; Principles and Applications. Cambridge University Press, 1987.

14. Moo-Young M. Fermenter Immobilized Enzymes and Cells; Fundamentals and Applications. New York. Elsevier, 1988.

15. Karel SF, Libicki SB and Robertson, R. The immobilization of whole cells. engineering principles. Chem Eng Sci 1985; 10:1321-1354.

16. Berkeley RCW, Lynch JM, Melling J, Rutter PR and Vincent B, eds. Microbial Adhesion to Surfaces. Chichester. Ellis Horwood, 1980.

17. Vega JL, Clausen EC and Gaddy JL. Biofilm reactors for ethanol production. Enzyme Microb Technol 1988; 10:390-402.

18. Butler M, Hassell T and Rowley A. The use of microcarriers in animal cell cultures. In: Webb C and Mavituna F, eds. Plant and Animal Cells; Process Possibilities. Chichester: Ellis Horwood, 1987:64-74.

19. Cooper PF and Atkinson B, eds. Biological Fluidised Bed Treatment of Water and Wastewater. Chichester: Ellis Horwood, 1981.

20. Kierstan MPJ and Coughlan MP. Immobilization of cells and en-

zymes by entrapment. In: Woodward J, ed. Immobilized Cells and Enzymes—A Practical Approach. Oxford: IRL Press, 1985:39-48.

21. Bucke C. Methods of immobilizing cells. In: Webb C, Black GM and Atkinson B, eds. Process Engineering Aspects of Immobilized Cell Systems. Rugby: IChemE Publ, 1986:20-34.

22. Anonymous. Protanal Alginates for Cell Immobilization. Information Sheet EF NO 1006/1, Norway: Protan, 1982.

23. Emery AN and Mitchell DA. Operational considerations in the use of immobilized cells. In: Webb C, Black GM and Atkinson B, eds. Process Engineering Aspects of Immobilized Cell Systems. Rugby: IChemE Publ, 1986:87-99.

24. Greenshields RN and Smith EL. Tower fermentation systems and their applications. The Chem Engr 1971; 249:182-190.

25. Metz B and Kossen NWF. The growth of moulds in the form of pellets—a literature review. Biotechnol Bioeng 1977; 19:781-799.

26. Winkler MA. Biological Treatment of Aqueous Wastes. Chichester: Ellis Horwood, 1981.

27. Prince IG and Barford JP. Induced flocculation of yeasts for use in the tower fermenter. Biotechnol Lett 1982; 4:621-626.

28. Rupp RG. Use of cellular microencapsulation in large scale production of monoclonal antibodies. In: Feder J and Tolbert WR, eds. Large Scale Mammalian Cell Culture. New York: Academic Press, 1985:19-36.

29. Chang HN. Membrane bioreactors: engineering aspects. Biotechnol Adv 1987; 5:129-145.

30. Feder J and Tolbert WR. Large Scale Mammalian Cell Culture. New York: Academic Press, 1985.

31. Bandyopadhyay K K and Ghose T K. Studies on immobilized *Saccharomyces cerevisiae*: III Physiology of growth and metabolism on various supports. Biotechnol Bioeng 1982; 24:805-815.

32. Navarro JM and Durand G. Modifications of the growth of *Saccharomyces uvarum* by immobilization on a solid carrier. C R Hebd Seances Acad Sci Ser D 1980; 290:453-456.

33. Doran PM and Bailey JE. Effects of immobilization on growth, fermentation properties, and macromolecular composition of *Saccharomyces cerevisiae* attached to gelatin. Biotechnol Bioeng 1986; 28:73-87.

34. Navarro JM and Durand G. Modification of yeast metabolism by immobilization onto porous glass. Eur J Appl Microbiol 1977; 4:243-254.

35. Durand G and Navarro JM. Use and properties of cells of immobilized microorganisms. Microbiol Ind Aliment Ann Congr Int 1980; 3:97-109.

36. Lindsey K and Yeoman MM. The viability and biosynthetic activity of cells of *Capsicum frutescens* Mill cv annuum immobilized in reticulate polyurethane. J Exp Bot 1984; 35:1684-1696.

37. Lindsey K and Yeoman MM. The synthetic potential of immobilized cells of *Capsicum frutescens* mill cv annuum. Planta 1984; 162:495-501.

38. Subramani J, Bhatt PN and Mehta AR. Alkaloid production by the immobilized cells of *Solanum xanthocarpum*. Curr Sci 1989; 58:510-521.

39. Asada M and Schuler ML. Stimulation of ajmalicine production and excretion from *Catharanthus roseus*: effects of adsorption in situ, elicitors and alginate immobilization. Appl Microbiol Biotechnol 1989; 30:475-481.

40. Von Felten P, Zuerrer H and Bachofen R. Production of molecular hydrogen with immobilized cells of *Rhodospirillum rubrum*. Appl Microbiol Biotechnol 1985; 23:15-20.

41. Kuwada Y and Ohta Y. Hydrogen production by an immobilized cyanobacterium *Lyngbya* sp. J Ferment Technol 1987; 65:597-602.

42. Callegari JP, Deplano A and Simon JP. Effect of immobilizing *Chlorella sorokiniana* on photosynthesis and excretion of organic compounds. World Biotech Rep 1985; 1:651-654.

43. Zhang X, Bury S, DiBiasio D and Miller JE. Effects of immobilization on growth, substrate consumption, β-galactosidase induction, and byproduct formation in *Eschericia coli*. J Ind Microbiol 1989; 4:239-246.

44. Chen J and Wu Y. The effects of immobilized micro-environment on the yeast metabolism. Shengwu Gongcheng Xuebao 1987; 3:146-151.

45. Evans PJ and Wang HY. Pigment production from immobilized *Monascus* sp utilizing polymeric resin adsorption. Appl Environ Microbiol 1984; 47:1323-1326.

46. Doran PM and Bailey JE. Effects of hydroxyurea on immobilized and suspended yeast fermentation rates and cell cycle operation. Biotechnol Bioeng 1986; 28:1814-1831.

47. Doran PM and Bailey JE. Effects of immobilization on the nature of glycolytic oscillations in yeast. Biotechnol Bioeng 1987; 29:892-897.

48. Galazzo JL, Shanks JV and Bailey JE. Comparison of suspended and immobilized yeast metabolism using phosphorus-31 nuclear magnetic resonance spectroscopy. Biotechnol Tech 1987; 1:1-6.

49. Galazzo JL and Bailey JE. In vivo nuclear magnetic resonance analysis of immobilization effects on glucose metabolism of yeast *Saccharomyces cerevisiae*. Biotechnol Bioeng 1989; 33:1283-1289.

50. Vijayalakshmi M, Marcipar A, Segard E and Broun GB. Matrix-bound transition metal for continuous fermentation tower packing. Ann N Y Acad Sci 1979; 326:249-254.

51. Mattiasson B and Hahn-Haegerdal B. Microenvironmental effects on metabolic behavior of immobilized cells: a hypothesis. Eur J Appl Microbiol Biotechnol 1982; 16:52-55.

52. Mattiasson B, Larsson M and Hahn-Haegerdal B. Metabolic behavior of immobilized cells—effects of some microenvironmental factors. Ann N Y Acad Sci 1984; 434:475-478.
53. Rosevear A and Lambe CA. Biological production of chemical compounds. Eur Pat Appl EP 62457 A2 13 Oct 1982.
54. Shuler ML. Immobilized whole cell bioreactors: potential tools for directing cellular metabolism. World Biotech Rep 1985; 2:231-239.
55. Haldimann D and Brodelius P. Redirecting cellular metabolism by immobilization of cultured plant cells: a model study with *Coffea arabica*. Phytochemistry 1987; 26:1431-1434.
56. Ayabe S, Iida K and Furuya T. Induction of stress metabolites in immobilized *Glycyrrhiza echinata* cultured cells. Plant Cell Rep 1986; 5:186-189.
57. Mahmoud W and Rehm HJ. Morphological examination of immobilized *Streptomyces aureofaciens* during chlortetracycline fermentation. Appl Microbiol Biotechnol 1986; 23:305-310.
58. Pertot E, Rozman D, Milicic S and Socic H. Morphological differentiation of immobilized *Claviceps paspali* mycelium during semicontinuous cultivation. Appl Microbiol Biotechnol 1988; 28:209-213.
59. Shirai Y, Hashimoto K, Yamaji H and Kawahara H. Oxygen uptake rate of immobilized growing hybridoma cells. Appl Microbiol Biotechnol 1988; 29:113-118.
60. Felix H, Brodelius P and Mosbach K. Enzyme activities of the primary and secondary metabolism of simultaneously permeabilized and immobilized plant cells. Anal Biochem 1981; 116:462-470.
61. Brodelius P and Nilsson K. Permeabilization of immobilized plant cells, resulting in release of intracellularly stored products with preserved cell viability. Eur J Appl Microbiol Biotechnol 1983; 17:275-280.
62. Beaumont MD and Knorr D. Effects of immobilizing agents and procedures on viability of cultured celery: *Apium graveolens* cells. Biotechnol Lett 1987; 9:377-382.
63. Knorr D and Teutonico RA. Chitosan immobilization and permeabilization of *Amaranthus tricolor* cells. J Agric Food Chem 1986; 34:96-97.
64. Knorr D and Berlin J. Effects of immobilization and permeabilization procedures on growth of *Chenopodium rubrum* cells and amarantin concentration. J Food Sci 1987; 52:1397-1400.
65. Knorr D, Beaumont MD and Pandya Y. Polysaccharide copolymers for the immobili zation of cultured plant cells. In: Hollo J and Torley D, eds. Biotechnology Food Industry. Budapest: Akad Kiado, 1988:389-400
66. Santos-Rosa F and Galvan F. Ammonium photoproduction by free and immobilized cells of *Chlamydomonas reinhardtii*. Appl Microbiol Biotechnol 1989; 31:55-58.

67. Santos-Rosa F, Galvan F and Vega JM. Biological viability of *Chlamydomonas reinhardtii* cells entrapped in alginate beads for ammonium photoproduction. J Biotechnol 1989; 9:209-219.

68. Fang BS, Fang HY, Wu CS and Pan CT. High productivity ethanol production by immobilized yeast cells. Biotechnol Bioeng Symp 1984; 13:457-464.

69. Vieira AM, Sa-Correia I, Novais JM and Cabral JMS. Could the improvements in the alcoholic fermentation of high glucose concentrations by yeast immobilization be explained by media supplementation? Biotechnol Lett 1989; 11:137-140.

70. Passarinho PCLV, Vieira AMS, Cabral JMS, Novais JM and Kennedy JF. Effect of carrier matrix on fermentative production of ethanol by surface immobilized yeast cells. J Chem Technol Biotechnol 1989; 44:183-194.

71. Webb C, Fukuda H and Atkinson B. The production of cellulase in a spouted bed fermentor using cells immobilized in biomass support particles. Biotechnol Bioeng, 1986; 28:41-50.

72. Kopp B and Rehm HJ. Alkaloid production by immobilized mycelia of *Claviceps purpurea*. Eur J Appl Microbiol Biotechnol 1983; 18:257-263.

73. Murdin AD, Thorpe JS, Groves DJ and Spier RE. Growth and metabolism of hybridomas immobilized in packed beds: comparison with static and suspension cultures. Enzyme Microb Technol 1989; 11:341-346.

74. Karel SF, Briasco CA and Robertson CR. The behavior of immobilized living cells Characterization using isotopic tracers. Ann N Y Acad Sci 1987; 506:84-105.

75. Fernandez EJ, Mancuso A and Clark DS. NMR spectroscopy studies of hybridoma metabolism in a simple membrane reactor. Biotechnol Prog 1988; 4:173-183.

76. Vogel HJ and Brodelius P. An in vivo phosphorus-31 NMR comparison of freely suspended and immobilized *Catharanthus roseus* plant cells. J Biotechnol 1984; 1:159-170.

77. Mueller W, Wehnert G and Scheper T. Fluorescence monitoring of immobilized microorganisms in cultures. Anal Chim Acta 1988; 213:47-53.

78. Dervakos GA and Webb C. On the merits of viable cell immobilization. Biotechnol Adv 1991; 9:559-612.

CHAPTER 2

IMPROVED BIOLOGICAL STABILITY

There is now considerable evidence that the operational stability of immobilized cells may be substantially higher than that of freely suspended cells. An often quoted example is the longevity of the metabolic activity of immobilized *Claviceps fusiformis* which was reported to be as high as 550 days though the cells underwent significant morphological changes during this period.[1] Enhanced stabilities have also been demonstrated on a large scale. A continuous pilot plant reactor for ethanol production from nonsterilized molasses by yeast cells immobilized in Ca-alginate beads which contained sterols and unsaturated fatty-acids was reported to operate for more than 180 days.[2]

Not only the operational, but also the storage stability of viable cells can be extended by immobilization (Example 2: Enhanced Stability in a Continuous Beer Fermenter). Mugnier and Jung[3] reported that the number of living cells remained constant for periods of storage greater than three years at 28°C when the inocula were kept at a water activity less than 0.069. The cell survival was related to some properties of the water resulting from its interactions with the biopolymer.

The improvement of biological stability through cell immobilization offers significant opportunities in the area of genetic engineering where the instability of cell cultures containing plasmid vectors poses a major obstacle in the commercial realization of molecular cloning techniques. McLoughlin[4] recently claimed:

"The creation of microniches in time and/or space can enhance plasmid stability. Transient operation based on defined heterogeneous or dynamic environments found in gel immobilized cultures have

resulted in enhanced stability. Spatial organization resulting from immobilization has the additional advantage of regulated cell protection within defined micro-environments and controlled release, depending on the nature of the gel, from these micro-environments or microcosms. This regulation of ecological competence allied to the advantages of microbial cell growth in gel microenvironments combined with the spatial organization (or juxtapositioning of cells, selective agents, nutrients, protectants, etc.) possible through immobilization technology offers new strategies to enhance plasmid and culture stability."

Apart from the obvious cost advantages, the increased stability of immobilized cells may lead to advantageous stoichiometric changes, such as increased product yields. This has been demonstrated repeatedly with immobilized plant cells. The enhanced alkaloid accumulation in calcium alginate entrapped cells of *Catharanthus roseus* using a limiting growth medium was attributed to the stabilization of the cells by the matrix.[5,6] Immobilized cells of *Solanum surattense* released far more solasodine into the medium than free cell suspension cultures; this enhancement was ascribed to the stabilization of cells after immobilization as well as the effect of growth hormones in the medium.[7] Caution, however, should be applied in ascribing increased product yields to the stabilization of the metabolic activities of the cell by immobilization, as a number of other factors, such as the reduced cell growth rate or metabolic effects, may play an important role.

MECHANISMS OF STABILIZATION

A number of hypotheses have been proposed in order to explain the often remarkable increases in the biological stability of immobilized cells (Table 2.1). It appears that the actual mechanism for this stabilization depends on the type and physiological state of the immobilized cells. When the cells are not viable then the enhanced catalytic stability can be best attributed to the protective effect of the immobilization matrix against physico-chemical challenges such as temperature, pH and organic solvents. With viable non-growing or slowly-growing cells, additional mechanisms become important such the increased biosynthesis of intracellular enzymes compensating for enzyme de-activation/de-naturation, the capability of the cells to grow on compounds excreted by lysed cells and the possibility of reduced cell mass degradation rates.

Table 2.1. Mechanisms of cell stabilization

1	Cell growth
2	Protection from physico chemical challenges
3	'Cryptic' growth
4	Possibility of reduced cell mass degradation
5	Reduced number of cell divisions
6	Diffusion limitations

When the cells are growing, then the maintenance of a balance–either fortuitously or intentionally–between cell growth and cell elusion/deactivation becomes very important in maintaining a constant metabolic activity over long periods of time.

The increased plasmid stability of genetically engineered immobilized cells in the absence of selection pressures has been ascribed to the decreased number of cell divisions within the matrix which does not allow plasmid-free cells to appear and/or take over the culture. Mechanical stabilization is important for shear sensitive cells such as animal cells which find protection within the low-shear environment of a porous matrix. In addition, immobilization seems to preserve the structure of the cells and the spatial positioning of enzyme sequences within the cells. The various mechanisms of cell stabilization are analyzed below.

CELL GROWTH

The capability of the cells to grow is very important in restoring the inevitable decline of the catalytic activity of living cells. However, the cell growth rate which gives the best stability is not necessarily the optimum for the metabolic activity of the cells. For example, when the cells are engaged in the production of non-growth associated metabolites, they will typically have a very low or nil growth rate which might be inadequate to compensate for cell mass degradation and protein deactivation. This will result in the decline and eventual cessation of the activity. On the other hand, when cells are required to have a relatively high growth rate, i.e. in growth-associated fermentations, then they will eventually reach the physico-mechanical limits imposed by the equipment with potential disastrous effects (rupture of carrier, clogging of the reactor, etc). An added complication is the spatial heterogeneity of cell growth within an immobilized cell particle brought about by diffusional limitations, which make the accurate control of cell growth very difficult.

In certain cases a balance between cell growth and cell decay/ leakage may be established fortuitously. An example is the continuous production of acetic acid by immobilized *Acetobacter aceti* cells entrapped in a κ-carrageenan gel, a highly growth associated process.[8] Although cells were gradually released into the medium resulting in a reduced number of viable cells in the gel, the productivity of the system remained constant for a period of 120 days. This was attributed to a combined action of both immobilized and free cells. In other cases such a balance may be maintained by computer control. For example, a constant penicillin-G concentration in the effluent stream of a fluidized bed reactor could be sustained for more than 25 days by controlling the cell specific growth rate at a desired value.[9]

The most common mode of operation reported to stabilize the activity for long periods of time, especially with non-growth associated products, is the periodic cycling of immobilized cells between growth and non-growth conditions. The forced oscillation of immobilized cell systems can be achieved by cycling an essential nutrient, cycling some other condition such as light,[10] by using a particular reactor configuration (e.g. the rotating disk contactor) or by simply using a repeated-batch operation (Example 3: Enhanced Stability Through Repeated Batch Operation: Ferrous Sulphate Oxidation).

Periodic operation of immobilized cells has been used in a number of cases. Foerberg et al[11,12] have shown that nitrogen-starved cells of *Clostridium acetobutylicum* immobilized in alginate could maintain butanol production for eight weeks when rejuvenated for 15 minutes every eight hours in a nitrogen-rich medium. Citric acid, which is produced optimally at very low nitrogen concentrations, is an obvious candidate for periodic operation and has been investigated by many researchers.[13-15]

Periodic operation of immobilized cells has also been applied to the production of a number of antibiotics. The longevity of patulin production was investigated in detail by Gaucher's group. It was reported that when *Penicillium urticae* cells were semi-continuously transferred to a nitrogen-free medium every 48 h, the half-life of patulin production was 16 days versus only 4-6 days for free cells. This half-life could be further improved by occasional or repeated incubations of the immobilized cells in dilute growth media.[16,17] This long-term production was the result of the formation of new cells rather than the maintenance of the original ones.[18] The lower half-lives of the free cells and of the immobilized cells under nitrogen starvation were attributed to the gradual loss of the cells' catalytic capacity for converting glucose to 6-methylsalicylic acid, the first metabolite of the patulin pathway.[19,20] The production of penicillin-G by *Penicillium chrysogenum* cells immobilized in κ-carrageenan beads exhibited a nine-fold longer half-life than that exhibited by free cells, when the cells were alternated between glucose-containing and glucose-free media.[21,22] The semicontinuous production of chlortetracycline by immobilized cells of *Streptomyces aureofaciens* showed a four-fold increase in the half-life compared with that of free cells.[23] Morikawa et al[24,25] reported that the bacitracin activity of immobilized cells of a *Bacillus* species was increased upon 13 repeated incubations in a peptone medium while the activity of washed cells dropped considerably after seven repeated incubations. Similar investigations have been conducted for a variety of other products such as oxytetracycline[26] actinomycin D,[27] glucoamylase,[28,29] acid proteinase[30] and α-peptide by recombinant yeast cells.[31]

In the production of growth-associated metabolites, stability depends on the successful de-coupling of cell growth from product formation which will minimize operational problems due to cell overgrowth. The experimental evidence so far, however, suggests that this may not always be possible. Wada et al[32] reported a steep drop in the ethanol productivity of immobilized yeast to values approaching zero after 2 to 4 days in nitrogen-free medium. Inloes et al[33] reported that the periodic perfusion of a hollow-fiber membrane bioreactor used in the production of ethanol by nitrogen-deficient yeast cells resulted in a decline of the fermentation efficiency with a concomitant reduction in the total protein concentration of immobilized cells within the hollow-fiber membranes. During nitrogen-deficiency the specific ethanol productivity dropped to less than 10% of that of the complex medium. In subsequent growth phases, ethanol production rates increased to levels 40-70% of the initial growth-phase values, but the ability to regenerate the fermentation activity decreased with culture age.

Amin et al[34] reported that immobilized growing cells were necessary for the long-term, continuous production of xylose by

Pachysolen tannophilus, since a steady state ethanol concentration was maintained for only one week by immobilized non-growing cells. Nevertheless, reactivation, especially under aerobic conditions, has been reported to enhance the stability of yeasts.[35-39] Similarly, attempts to restrict the growth rate of immobilized cells but maintain specific production rate were unsuccessful for *Caldariomyces fumago* in the production of chloroperoxidase[40] and *Clostridium beijerinckii* in the isopropanol-butanol-ethanol fermentation.[41]

PROTECTION FROM PHYSICO-CHEMICAL CHALLENGES

The increased stability of intracellular activities can be attributed largely to the protective effect of the immobilization matrix against physicochemical challenges (temperature, pH, solvents, shear, heavy metals) which, for example, may de-activate various intracellular enzymes. Similarly, strictly anaerobic cells embedded within a matrix may find protection from the toxicity of oxygen.

The range of pH over which *Kluyveromyces marxianus* and *Zymomonas mobilis* exhibit the highest specific ethanol productivities was reported to be broader in comparison to freely suspended cells, though the underlying mechanism was not identified.[42,43]

The protection of immobilized cells from the denaturating effects of organic solvents has been demonstrated in the extractive fermentation of ethanol; the degree of protection, however, appears to depend on the nature of the solvent. *Saccharomyces bayanus* was reported to be protected from the toxic effects of oleic acid when immobilized in κ-carrageenan.[44] Similarly, Porapack Q was found to be an effective barrier against solvents.[45] The resistance of immobilized cells to solvent de-activation may be further improved by proper design of the immobilization carrier. Addition of vegetable oils, which are absorbents of solvents, to Ca-alginate enhanced further the catalytic stability of immobilized cells against a wider spectrum of toxic solvents.[46,47] Shueler et al[48] reported that the stability of immobilized *Pseudomonas oleovorans* cells producing 1,2-epoxyoctane from 1-octene, as evidenced by the oxygen consumption rate, was considerably greater than that of free cells and this was attributed to the immobilization matrix preventing direct contact between the cells and the 1-octene.

The mechanical stabilization of the cells by the immobilization matrix has been reported to improve the stability of shear sensitive cells such as animal cells. A recent study, however, demonstrated that shear stress in certain ranges may not be detrimental to mammalian cells. In fact, throughout the range of shear stresses studied, the metabolite production was maximized by maximizing shear stress.[49]

CRYPTIC GROWTH

The ability of immobilized cells to assimilate products released from lysed cells for growth/maintenance purposes is important in maintaining a constant catalytic activity, especially when the reaction media do not allow growth of cells, e.g. in biotransformations with viable cells. For example, Koshcheenko et al[50,51] concluded that cryptogenic growth in cells of *Arthrobacter globiformis*, *S. cerevisiae* and *Bacillus megatherium* applied in the transformation of steroids is important in determining the high activity and stability of cells in the immobilized state. In another report[52] the chemoautotroph *Thiobacillus denitrificans* was immobilized in macroscopic floc by co-culture with floc-forming heterotrophs; no organic carbon addition was required during enrichment for immobilized cells of *T. denitrificans* or H_2S oxidation. This was attributed to the heterotrophs deriving organic carbon for maintenance and growth from waste products of *T. denitrificans* and products of cell lysis.

CELL MASS DEGRADATION RATE

The effect of immobilization on the rate of cell mass degradation–which is often considered to be a first order rate process–has not been studied thoroughly so far. Based on published information, it appears that there may not be significant differences in the cell degradation rate constants for immobilized and free cells. Toda and Sato[53] estimated the death rate constant of immobilized *Candida lipolytica* to be 45×10^{-3} h^{-1}, which is roughly half the value of the intact cells. However, Karel and Robertson[54,55] used pulse-chase radioisotope labeling with $^{35}SO_4^{2-}$ to estimate cell mass degradation for *Pseudomonas putida* and *E. coli*. From the curve of sulfur release it was estimated that their respective long-term rate decay constants were $1\text{-}4 \times 10^{-3}$ h^{-1} and 1×10^{-3} h^{-1} which compared very well with the values for aerobic and anaerobic carbon starvation for free cells.

REDUCED NUMBER OF CELL DIVISIONS

The mechanical properties of certain immobilized cell systems may allow only a limited number of cell divisions to occur; this may contribute to the enhanced stability of cells which would otherwise become unstable after a number of cell divisions. For example, with some genetically engineered cells a number of generations are required before plasmid-free segregants begin to appear.

The enhanced stability of the recombinant plasmid pTG201 containing the *xylE* gene in immobilized *E. coli* cells in the absence of antibiotic selection has been demonstrated by many investigators.[56-59] De Taxis du Poet et al[56] attributed this stability to the compartmentalization resulting from the immobilized growth conditions: the mechanical properties of a gel-bead system allow only a limited number of cell divisions (10-16) to occur in each clone of cells before the clone escapes from the gel bead. This number of generations is not sufficient for plasmid-free cells to appear within the cavities (plasmid-free segregants appeared to have a lag period of 25-30 generations). Even when they appear, they cannot overcome the culture.[58] Besides the mechanical properties of gel beads which allow only a limited number of cell divisions, an increase in the plasmid copy number seems to be involved in the enhanced plasmid stability in immobilized cells.[57,59-62]

Enhanced stability for genetically engineered cells has also been reported in a number of other applications. The contribution of immobilized cells to plasmid stability was studied with recombinant *E. coli* containing plasmids with the *lacZ* gene. Plasmid stability was much higher in the immobilized cells, with 44% retention after 125 reactor dilutions compared to 5% retention after 100 reactor dilutions with the free cells.[63] Relatively high stabilities were also observed in the production of human chorionic gonadotropin α-subunit by genetically engineered *S. cerevisiae*,[64] in ethanol fermentations with immobilized yeast containing killer plasmids[65] and in β-lactamase production by *Streptomyces lividans*.[66] The continuous selective excretion of a recombinant protein by immobilized *E. coli* was reported to be stable for 120 days.[67] The feasibility of the long-term, high-level continuous production of the growth hormone somatomedin C with immobilized transformed yeast cells was demonstrated by Sode et al.[68]

DIFFUSION LIMITATIONS

Diffusional artefacts may also be responsible for the apparent longevity of immobilized cells. When an immobilized cell particle is diffusion-limited and there is no cell growth then a decline in the activity due to enzymic decay/cell elusion may not become immediately apparent because of the concomitant increase in the effectiveness factor.

BUT WHAT ABOUT PARTICLE STABILITY?

While immobilization may improve considerably the biological stability of the cells, the overall stability of the system may be compromised by the poor stability of the immobilization particles themselves. This is particularly so for gel particles exposed to the high-shear environment of a stirred tank reactor (STR). However, BSPs appear to provide more robust performance, even in the harsh environment of a STR (Example 4: Particle Stability in a Stirred Tank Reactor).

EXAMPLE 2:
ENHANCED STABILITY IN A CONTINUOUS BEER FERMENTER

Ideas for improving the batch process of beer production date back to the end of the nineteenth century. These ideas were, however, largely impractical owing to the primitive methods of vessel design and construction and to failure through microbial contamination during extended operation. A reawakening of interest in rapid beer fermentation occurred in about 1955. Due to the general increase in beer consumption all over the world, the demand for shortening processing times in breweries became greater. This also stimulated a great deal of research into rapid beer fermentation and consequently, theoretical and practical advances important in continuous fermentation were established. Anticipated advantages from rapid beer production included: reduced capital costs, less capital tied-up as beer in the process, reduced labor costs and lower product costs (F.C. Yang, PhD Thesis, UMIST, 1989).

Before techniques of cell immobilization were introduced around 1970, most novel fermentation systems relied on the sedimentation of yeast to build up high yeast concentration. The methods used in these systems include: recycling the yeast cells collected from the settling vessel (multi-vessel systems); selecting very flocculent yeast strains (tower fermenter systems); and using a special facility or design to enhance cell settling (inclined-tube fermenters, centrifugal fermenters). Although many alternative processes for beer production have been proposed, most of them have been unsuccessful and have not progressed beyond the laboratory scale. Only two systems, the multi-vessel and the tower fermenter, have been commercialized successfully, and even these systems suffer drawbacks which prevent them from being widely accepted.

In this example a promising alternative beer production system based on BSPs is reported. After developing BSP colonization techniques (Example 7: Colonization of BSPs by Filtration, chapter 3) fermentation tests, using yeast cells immobilized in BSP-packed bed bioreactor, were carried out.

MATERIALS AND METHODS

The source of yeast cells and the preparation of yeast slurries for this example are described in chapter 3 (Example 7: Colonization of BSPs by Filtration). Before being used for BSP colonization and beer fermentation, the yeast slurries were kept aseptically in a modified freezer at 0°C and also checked at intervals for viability and contamination.

The bioreactor consisted mainly of a QVF glass jacketed column (350 mm x 50 mm ID) within which the fermentation temperature could be controlled by circulating water through the jacket. Both end sections of the bioreactor were made of perspex tube (50 mm x 50 mm ID) and plates (10 mm thick). Each end section was designed to have three inlet/outlet ports for wort and CO_2 so that the bioreactor could be operated in different positions. The three sections of the bioreactor were held together by two sets of QVF flanges and sealed with neoprene rubber gaskets. Perforated stainless-steel plates were used for retaining the BSPs. An outline of the experimental setup is described in Example Figure 2.1.

When the BSP colonization was completed, the bioreactor was fixed in a vertical, horizontal or inclined position. The brewing wort was pumped into the bottom part of the bioreactor continuously and effluent (beer) from the bioreactor flowed directly into a modified freezer, in the side of which a small hole had been drilled for passage and within which the temperature was controlled at very slightly above 0°C. At this temperature yeast activity is very much reduced but since cells or product were not actually frozen, damage to viability or quality was avoided. Silicone rubber tubes were employed for connecting the

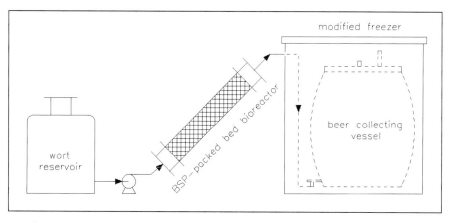

Example Fig. 2.1. Inclined packed bed bioreactor.

whole system. The effluent was sampled at regular intervals and specific gravity (SG), pH, VDK and yeast concentration were measured. At the end of each experiment BSP biomass content at different positions within the bioreactor was also determined.

Beer produced from fermentation tests was first cooled in the freezer, decanted from sedimented yeasts and then dispensed cold to bottles which were then capped with crown tops using a two lever crown corker (Boots Beermaker). The bottled beer was stored at 0°C for taste tests.

After a certain period of continuous fermentation the feed system was switched off and the colonized BSP-packed bioreactor was separated from the feed system and the product collecting vessel. All the tubes connected with the bioreactor ports were clamped and plugged with cotton wool and capped with aluminum foil. The bioreactor was then removed into the 0°C modified freezer for storage. When the fermentation operation was to be restored, the bioreactor was taken from the freezer and the whole system built up again. The fermentation process was then continued by feeding wort into the colonized bioreactor. The properties of the effluent (beer) were determined at regular intervals.

RESULTS AND DISCUSSION

After a month of storage the BSP-packed bed bioreactor was reused and the results are shown in Example Figure 2.2. It was observed that when a brewing wort with a specific gravity (SG) of 1040 was fed into the bioreactor, yeast cells were reactivated very rapidly. The fermentation reaction occurred immediately and carbon dioxide was also generated. As can be seen in Example Figure 2.2, the SG of the product

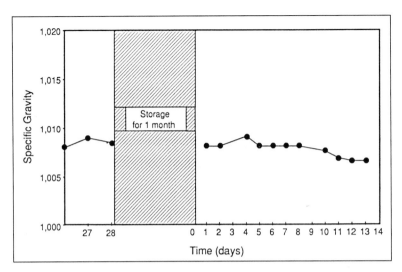

Example Fig. 2.2. The stability of an immobilized cell fermenter for the continuous production of beer, after an interruption of operation of one month.

dropped to less than 1010 within one day and after two days of continuous fermentation, the bioreactor reached steady state once more. The properties and the flavor of the product were found to be very similar to those of the beer produced before the storage.

Early work showed that yeast cells can be stored in beer or water for a few weeks before use. The results presented here for the colonized BSP-packed bed bioreactor show that they can also be removed from continuous operation, stored over a few weeks and then reused for beer fermentation.

CONCLUSION

The continuous fermentation process based on yeast cells immobilized in BSPs can be stopped temporarily and, following storage under suitable conditions, can be restarted without any need for re-colonization. This is an extreme example of feed flowrate fluctuation and demonstrates the stability of the immobilized cell system.

ACKNOWLEDGMENT

The material for this example was adapted from the PhD thesis of Dr. Fan-Chiang Yang (UMIST, 1989).

EXAMPLE 3:
ENHANCED STABILITY THROUGH REPEATED BATCH OPERATION: FERROUS SULPHATE OXIDATION

Biological oxidation of ferrous sulphate $FeSO_4$ using *Thiobacillus ferrooxidans* is a well-researched area in biohydrometallurgy in the treatment of acid mine drainage. It has also recently been proposed as a means of treating hydrogen-sulphide-bearing gases and for purifying wastewaters. The natural tendency of *T. ferrooxidans* to grow on surfaces makes it an ideal organism for cell immobilization and a number of workers have exploited this tendency in attempts to increase ferrous sulphate oxidation rates. Various immobilization methods have been employed, including the use of ion exchange resins, glass beads, activated carbon particles, PVC and diatomaceous earth. In mineral leaching processes the immobilization support is the sulphide ore which is itself oxidized by the bacteria attached to it.

In this example immobilization of *T. ferrooxidans* cells in polyurethane foam BSPs was used as a basis for repeated batch experiments.

MATERIALS AND METHODS

A freeze-dried culture of *T. ferrooxidans* was obtained from the National Collection of Industrial and Marine Bacteria (Aberdeen, UK)(code no. NCIMB 9490 or ATCC 19859). The organism was grown and maintained on 9K medium. Cubic BSPs of 6 mm side length, having a porosity of around 0.97 and a pore size of 80 pores per linear inch (ppi) were used to immobilize the bacterial cells (H. Armentia, MSc Dissertation, UMIST, 1991).

Once immobilization was proven in shaken flasks an immobilized cell reactor was constructed for repeated batches and continuous oxidation trials (Example Fig. 3.1). The reactor consisted of a glass column (30.6 cm x 5.2 cm) with an inlet for medium at the top and an outlet for effluent at the bottom. Prior to autoclaving, 400 BSPs were placed in the reactor. A working volume of either 400 mL or 250 mL and an aeration rate of 400 mL/min were used for all experiments.

Eight consecutive batches were run using the immobilized reactor on a "draw and fill" basis, without any intermediate inoculation. At the end of the eighth batch, by which time steady state biomass levels had been achieved, the reactor was switched to continuous operation.

RESULTS AND DISCUSSION

Example Figure 3.2 shows results for the repeated batch oxidation of ferrous sulphate. The 40 hours required for complete oxidation in the

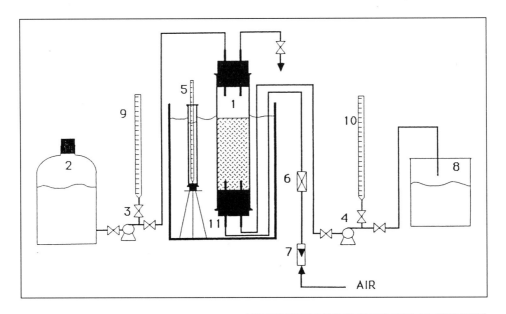

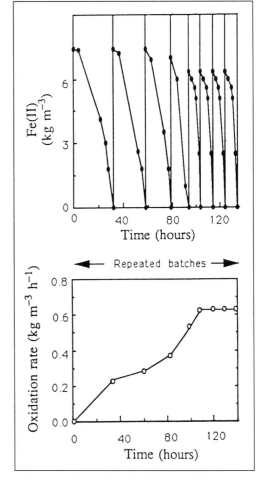

Example Fig. 3.1. (above) *Bioreactor system for immobilized* T. ferrooxidans. *1- circulating bed; 2- medium storage bottle; 3- medium feed pump; 4- liquid effluent pump; 5- thermometer; 6- air filter; 7- rotameter; 8- liquid effluent pump; 9- medium flowrate measuring pipette; 10- effluent flow rate measuring pipette; 11- water bath.*

Example Fig. 3.2. (right) *Results from repeated batch fermentations to oxidize FeSO₄ solutions using immobilized* Thiobacillus ferrooxidans *cells.*

first batch is exactly the same as it would be for freely suspended cells. However, it can be seen quite clearly that by immobilizing the cells and enabling their re-use, the length of time required to achieve full oxidation is reduced considerably after the first batch. Maximum productivity is not reached until the sixth batch, by which time it is three times higher than for the conventional single batch system.

The above results clearly show that cells continue to accumulate within BSPs, from batch to batch, so that volumetric biomass holdups and consequently productivities are increased beyond the levels normally achieved in a single operation. More importantly, the exposure of immobilized cells to a periodic oscillation of substrate concentration through the repeated batch operation, appears to significantly stabilize their behavior.

In a separate experiment, the operational stability of the reactor was tested by stopping operation for periods of 20 h, with liquid being drained from the column. It was found that the reactor operation could be immediately re-started after such interruption, simply by filling the column with medium.

The immobilized biomass in the foam particles clearly retained their oxidative efficiency during storage in the empty column. Indeed, even particles that had been removed from the column and stored in the open air for periods of up to 6 weeks were able to retain their activity and were used to restart reactor operations, although unfortunately no data were recorded for these tests. These results have implications for industrial operation since reactor operation could be stopped and re-started whenever required without further manipulation. With cells immobilized in BSPs the possibility also exists for long-term storage of viable cells outside the reactor environment.

CONCLUSION

The ability of the immobilized cell reactor to be stopped and re-started and to cope with long process interruptions makes it a suitable candidate for industrial applications where process stability and predictability is as important as process productivity.

ACKNOWLEDGMENT

The material for this example was adapted from the MSc Dissertation of Helene Armentia (UMIST, 1991).

EXAMPLE 4:
PARTICLE STABILITY IN A STIRRED TANK REACTOR

Stirred tank reactors (STRs) are widely used in fermentation plants not only because they are easy to operate and the know-how for their manufacture and operation is well established, but also because they generally provide suitable mass transfer conditions for most processes. They are not used, however, in fermentations carried out with immobilized cells, for which fluidized bed, fixed bed or even circulating bed reactors are more widely used, despite the fact that the mass transfer ability of such reactor types is frequently lower than that of the stirred tank. Despite the advantages of using immobilized cells, their association with reactors other than the 'state of the art' stirred tank represents a major obstacle to their use in industry in replacing traditional fermentation processes, especially in already established plants. They are, therefore, regarded as a very expensive process improvement or as a new (and therefore risky) alternative.

The reasons for not using the STR in fermentations carried out with immobilized cells are not clear. Extensive aggregate erosion caused by the high shear stress in the reactor may be a probable motive, but most of the tests reported in the literature appear to have been carried out with Ca-alginate beads and other gels, which do not have a high mechanical stability anyway. On the other hand, authors who have reported the use of more resistant immobilization supports, such as BSPs in stirred tanks, do not mention the extent of erosion. In this example, the effects of shear stress on the mechanical stability of immobilized cell particles in a STR is examined.

MATERIALS AND METHODS

Studies were carried out, at different particle holdups and aeration rates, in a 10 L stirred tank (P.A.L. Rodrigues, PhD Thesis, UMIST, 1995). This was 217 mm in diameter and was equipped with a 6-bladed Rushton turbine of 80 mm diameter, driven by an overhead motor and with 4 baffles placed at 90º with respect to each other. Tap water was used in all the tests. The support particles used were ca. 6 mm polyurethane foam BSPs with 60 pores per inch (ppi). The erosion suffered by the particles after being subjected to various rotation speeds and aeration rates was quantified by measuring the side length of the used particles.

RESULTS AND DISCUSSION

During the tests the particles stayed in the tank for about 4 days and were subjected during that time to impeller tip speeds that varied from

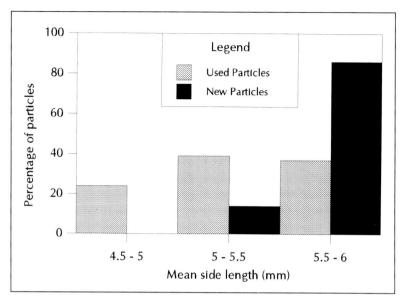

Example Fig. 4.1. Erosion suffered by BSPs after being subjected to various rotation speeds and aeration rates for 4 days.

0.72 to 2.88 m/s and aeration rates of 0.3 to 1.7 vvm. After that time, 100 particles were measured. The results in terms of mean side length were compared with those of 100 unused particles and are presented in Example Figure 4.1. As can be seen, there was a decrease in the average particle size and also a broadening of the particle size distribution. The decrease was, on average, 10%. This was less than expected, and was considered to be acceptable since the particles were not occupied by cells and were, therefore, more susceptible to erosion. A penicillin fermentation carried out with *Penicillium chrysogenum* immobilized onto the same type of particles in a 2 L STR for 7 days at impeller tip speeds of 0.91 - 1.84 m/s and aeration rates of 0.8 - 1 vvm showed less erosion.

CONCLUSION

The results show that polyurethane foam particles are not extensively damaged during periods of operation normally used in microbial fermentations. It is likely that effects will be even less in larger scale stirred tanks where the relative size of particles will be smaller. It should be pointed out, however, that particle stability is the exception rather than the rule in immobilized cell reactors. BSPs provide one of the few exceptions.

ACKNOWLEDGMENT

The material for this example was adapted from the PhD Thesis of Dr. Paulo A.L. Rodrigues (UMIST, 1995).

REFERENCES

 1. Kren V, Ludvik J, Kofronova O, Kozova J and Rehacek Z. Physiological activity of immobilized cells of *Claviceps fusiformis* during long-term semicontinuous cultivation. Appl Microbiol Biotechnol 1987; 26:219-226.
 2. Nagashima M, Azuma M, Noguchi S, Inuzuka K and Samejima H. Continuous ethanol fermentation using immobilized yeast cells. Biotechnol Bioeng 1984; 26:992-997.
 3. Mugnier J and Jung G. Survival of bacteria and fungi in relation to water activity and the solvent properties of water in biopolymer gels. Appl Environ Microbiol 1985; 50:108-114.
 4. McLoughlin AJ. Plasmid stability and ecological competence in recombinant cultures. Biotechnology Advances 1994; 12:279-324.
 5. Majerus F and Pareilleux A. Alkaloid accumulation in calcium alginate entrapped cells of *Catharanthus roseus* using a limiting growth medium. Plant Cell Rep 1986; 5:302-305.
 6. Pareilleux A and Majerus F. Continuous production of indole alkaloids by gel-entrapped cells of *Catharanthus roseus*. NATO ASI Ser A 1987; 1341:495-501.
 7. Barnabas NJ and David SB. Solasodine production by immobilized cells and suspension cultures of *Solanum surattense*. Biotechnol Lett 1988; 10:593-596.
 8. Osuga J, Mori A and Kato J. Acetic acid production by immobilized *Acetobacter aceti* cells entrapped in a κ-carrageenan gel. J Ferment Technol 1984; 62:139-149.
 9. Kalogerakis N, Linardos T, Behie LA, Svrcek WY and Gaucher GM. Computer control of continuous immobilized cell penicillin fermentations: simulation studies. Can J Chem Eng 1986; 64:581-587.
10. Brouers M, Collard F, Jeanfils J and Loudeche R. Long-term stabilization of photobiological activities of immobilized algae—photoproduction of hydrogen by immobilized adapted *Scenedesmus* cells. Sol Energy R&D Eur Community Ser D 1983; 171-178.
11. Foerberg C, Enfors SO and Haeggstroem L. Control of immobilized non-growing cells for continuous production of metabolites. Eur J Appl Microbiol Biotechnol 1983; 17:143-147.
12. Haeggstroem L and Foerberg C. Long-term stability of nongrowing immobilized cells of *Clostridium acetobutylicum* controlled by the intermittent nutrient dosing technique. Methods Enzymol 1988; 137:626-632.
13. Anderson JG, Blain JA, Divers M and Todd JR. Use of the disc fermenter to examine production of citric acid by *Aspergillus niger*. Biotechnol Lett 1980; 2:99-104.
14. Eikmeier H and Rehm HJ. Semicontinuous and continuous production of citric acid with immobilized cells of *Aspergillus niger*. Z Naturforsch 1987; 42:408-413.

15. Horitsu H, Adachi S, Takahashi Y, Kawai K and Kawano Y. Production of citric acid by *Aspergillus niger* immobilized in polyacrylamide gels. Appl Microbiol Biotechnol 1985; 22:8-13.
16. Deo YM, Costerton JW and Gaucher GM. Semi-continuous and continuous production of antibiotics by immobilized fungal cells Dev Ind Microbiol 1984; 25:491-503.
17. Jones A, Berk D, Lesser BH, Behie LA and Gaucher GM. Continuous production of patulin by immobilized cells of *Penicillium urticae* in a stirred tank reactor. Biotechnol Lett 1983; 5:785-790.
18. Berk D, Behie LA, Jones A, Lesser BH and Gaucher GM. The production of the antibiotic patulin in a three phase fluidized bed reactor. II. Longevity of the biocatalyst. Can J Chem Eng 1984; 62:120-124.
19. Deo YM and Gaucher GM. Effect of nitrogen supplementation on the longevity of antibiotic production by immobilized cells of *Penicillium urticae*. Appl Microbiol Biotechnol 1985; 21:220-227.
20. Gaucher GM, Lam KS, Grootwassink JWD, Neway J and Deo YM. The initiation and longevity of patulin biosynthesis. Dev Ind Microbiol 1981; 22:219-2322.
21. Deo YM and Gaucher GM. Semi-continuous production of the antibiotic patulin by immobilized cells of *Penicillium urticae*. Biotechnol Lett 1983; 5:125-130.
22. Deo YM and Gaucher GM. Semicontinuous and continuous production of penicillin-G by *Penicillium chrysogenum* cells immobilized in κ-carrageenan beads. Biotechnol Bioeng 1984; 26:285-295.
23. Mahmoud W and Rehm HJ. Chlortetracycline production with immobilized *Streptomyces aureofaciens*. II. Semicontinuous and continuous fermentation. Appl Microbiol Biotechnol 1987; 26:338-341.
24. Morikawa Y, Ochiai K, Karube I and Suzuki S. Bacitracin production by whole cells immobilized in polyacrylamide gel. Antimicrob Agents Chemother 1979; 15:126-130.
25. Morikawa Y, Karube I and Suzuki S. Continuous production of bacitracin by immobilized living whole cells of *Bacillus* sp. Biotechnol Bioeng 1980; 22:1015-1023.
26. Ogaki M, Sonomoto K, Nakajima H and Tanaka A. Continuous production of oxytetracycline by immobilized growing *Streptomyces rimosus* cells. Appl Microbiol Biotechnol 1986; 24:6-11.
27. Dalili M and Chau PC. Production of actinomycin D with immobilized *Streptomyces parvullus* under nitrogen and carbon starvation conditions. Biotechnol Lett 1988; 10:331-336.
28. Federici F, Miller MW and Petruccioli M. Glucoamylase production by immobilized *Aureobasidium pullulans* in sequential batch processes. Ann Microbiol Enzimol 1987; 37:17-24.
29. Guo Y, Huang L, Zhao M, Chen H, Peng X, Luo G and Li L. Production of glucoamylase by immobilized cells. II. Conditions for batch fermentation and semicontinous fermentation. Shipin Yu Fajiao Gongye 1988; 28-31.

30. Aleksieva P, Mikhailova L and Ivanova L. Biosynthesis of acid proteinase from free and immobilized cells of *Humicola lutea*. Dokl Bolg Akad Nauk 1988; 41:67-70.

31. Sode K, Morita T, Peterhans A, Meussdoerffer F, Mosbach K and Karube I. Continuous production of α-peptide using immobilized recombinant yeast cells. A model for continuous production of foreign peptide by recombinant yeast. J Biotechnol 1988; 8:113-122.

32. Wada M, Kato J and Chibata I. Continuous production of ethanol using immobilized growing yeast cells. Eur J Appl Microbiol Biotechnol 1980; 10:275-287.

33. Inloes DS, Michaels AS, Robertson CR and Matin A. Ethanol production by nitrogen-deficient yeast cells immobilized in a hollow-fiber membrane bioreactor. Appl Microbiol Biotechnol 1985; 23:85-91.

34. Amin G, Khallafalla G and Doelle HW. Comparative study of D-xylose conversion to ethanol by immobilized growing or non-growing cells of the yeast *Pachysolen tannophilus*. Appl Microbiol Biotechnol 1988; 27:325-332.

35. Cho GH, Choi CY, Choi YD and Han MH. Ethanol production by immobilized yeast and its CO_2 gas effects in a packed bed reactor" J Chem Technol Biotechnol 1982; 32:959-96.

36. Ichimura K, Watanabe A, Mishima K and Endo K. Ethanol production by yeast immobilized on stilbazolium group-containing polyvinyl alcohol. Jpn Kokai Tokkyo Koho JP 61/260870 A2 1986.

37. Lee CW and Chang HN. Examination of immobilized cells in a rotating packed drum reactor for the production of ethanol from D-glucose. Enzyme Microb Technol 1985; 7:561-563.

38. Lee TH, Ahn JC and Ryu DDY. Performance of an immobilized yeast reactor system for ethanol production. Enzyme Microb Technol 1983; 5:41-45.

39. Varma R, Chattopadhyay SK, Baliga BA, Srivastava A, Ghosh BK and Karanth NG. Immobilized biocatalytic reactor for continuous ethanol production: enhancement of life through operational strategies. J Microb Biotechnol 1986; 1:35-40.

40. Carmichael RD, Jones A and Pickard MA. Semicontinuous and continuous production of chloroperoxidase by *Caldariomyces fumago* immobilized in κ-carrageenan. Appl Environ Microbiol 1986; 51:276-280.

41. Krouwel PG, Groot W J, Kossen NWF and Van der Laan WFM. Continuous isopropanol-butanol-ethanol fermentation by immobilized *Clostridium beijerinckii* cells in a packed bed fermenter. Enzyme Microb Technol 1983; 5:46-54.

42. Bajpai P and Margaritis A. Effect of temperature and pH on immobilized *Zymomonas mobilis* for continuous production of ethanol. Biotechnol Bioeng 1986; 28:824-828.

43. Bajpai P and Margaritis A. The effect of temperature and pH on ethanol production by free and immobilized cells of *Kluyveromyces*

marxianus grown on Jerusalem artichoke extract. Biotechnol Bioeng 1987; 30:306-313.

44. Barros MR, Aires Cabral JMS and Novais JM. Production of ethanol by immobilized *Saccharomyces bayanus* in an extractive fermentation system. Biotechnol Bioeng 1987; 29:1097-1104.

45. Matsumura M and Maerkl H. Application of solvent extraction to ethanol fermentation. Appl Microbiol Biotechnol 1984; 20:371-377.

46. Honda H, Taya M and Kobayashi T. Ethanol fermentation associated with solvent extraction using immobilized growing cells of *Saccharomyces cerevisiae* and its lactose-fermentable fusant. J Chem Eng Jpn 1986; 19:268-273.

47. Kobayashi T, Taya M and Kawase M. Extractive fermentation with immobilized microbial cells Jpn Kokai Tokyo Koho JP 61/192291 A2 1986.

48. Schueller C, Van der Meer AB, Van Lelyveld PH and Joosten GEH. Long-term stability of immobilized *Pseudomonas oleovorans* cells in the production of fine chemicals. Prog Ind Microbiol 1984; 20:85-92.

49. Frangos JA, McIntire LV and Eskin SG. Shear stress induced stimulation of mammalian cell metabolism. Biotechnol Bioeng 1988; 32:1053-1060.

50. Koshcheenko KA, Sukhodol'skaya GV, Tyurin VS and Skryabin GK. Physiological biochemical and morphological changes in immobilized cells during repeated periodical hydrocortisone transformations. Eur J Appl Microbiol Biotechnol 1981; 12:161-169.

51. Koshcheenko KA, Turkina MV and Skryabin GK. Immobilization of living microbial cells and their application for steroid transformations. Enzyme Microb Technol 1983; 5:15-21.

52. Ongcharit C, Dauben P and Sublette K L. Immobilization of an autotrophic bacterium by coculture with floc-forming heterotrophs. Biotechnol Bioeng 1989; 33:1077-1080.

53. Toda K and Sato K. Simulation study on oxygen uptake rate of immobilized growing microorganisms. J Ferment Technol 1985; 63:251-258.

54. Karel SF and Robertson CR. Cell mass synthesis and degradation by immobilized *Escherichia coli*. Biotechnol Bioeng 1989; 34:337-356.

55. Karel SF and Robertson CR. Autoradiographic determination of mass-transfer limitations in immobilized cell reactors. Biotechnol Bioeng 1989; 34:320-336.

56. De Taxis du Poet P, Dhulster P, Barbotin JN and Thomas D. Plasmid inheritability and biomass production: comparison between free and immobilized cell cultures of *Escherichia coli* BZ18(pTG201) without selection pressure. J Bacteriol 1986; 165:871-877.

57. De Taxis du Poet P, Arcand Y, Bernier R Jr, Barbotin JN and Thomas D. Plasmid stability in immobilized and free recombinant *Escherichia coli* JM105(pKK223-200): importance of oxygen diffu-

sion growth rate and plasmid copy number. Appl Environ Microbiol 1987; 53:1548-1555.

58. Nasri M, Sayadi S, Barbotin JN and Thomas D. The use of the immobilization of whole living cells to increase stability of recombinant plasmids in *Escherichia coli*. J Biotechnol 1987; 6:147-157.

59. Sayadi S, Nasri M, Barbotin JN and Thomas D. Effect of environmental growth conditions on plasmid stability plasmid copy number and catechol 23-dioxygenase activity in free and immobilized *Escherichia coli* cells. Biotechnol Bioeng 1989; 33:801-808.

60. Berry F, Sayadi S, Nasri M, Barbotin JN and Thomas D. Effect of growing conditions of recombinant *E. coli* in carrageenan gel beads upon biomass production and plasmid stability. Biotechnol Lett 1988; 10:619-624.

61. Marin-Iniesta F, Nasri M, Dhulster P, Barbotin JN and Thomas D. Influence of oxygen supply on the stability of recombinant plasmid *pTG201* in immobilized E. coli cells. Appl Microbiol Biotechnol 1988; 28:455-462.

62. Sayadi S, Nasri M, Berry F, Barbotin JN and Thomas D. Effect of temperature on the stability of plasmid *pTG201* and productivity of xylE gene product in recombinant *Escherichia coli*: development of a two-stage chemostat with free and immobilized cells. J Gen Microbiol 1987; 133:1901-1908.

63. Bailey K, Vieth WR and Chotani GK. Analysis of bioreactors containing immobilized recombinant cells. Ann NY Acad Sci 1987; 506:196-207.

64. Karkare SB, Burke DH, Dean RC Jr, Lemontt J, Souw P and Venkatasubramanian K. Design and operating strategies for immobilized living cell reactor systems. Part II. Production of hormones by recombinant organisms. Ann NY Acad Sci 1986; 469:91-96.

65. Yamamoto T, Yagiu J, Ohta K, Hamano M, Ouchi K and Nishiya T. Breeding of an alcohol yeast with K2 type of killer plasmids and its application to continuous alcohol fermentation. Nippon Nogei Kagaku Kaishi 1984; 58:559-566.

66. Lenzini MV, Erpicum T, Frere JM, Dusart J, Nojima S and Ogawara H. Cloning and expression of the β-lactamase gene of *Streptomyces cacaoi* in *Streptomyces lividans*: enzyme production by immobilized cells. Symp Biol Hung 1986; 32:385-387.

67. Georgiou G, Chalmers JJ, Shuler ML and Wilson DB. Continuous immobilized recombinant protein production from *E. coli* capable of selective protein excretion: a feasibility study. Biotechnol Prog 1985; 1:75-79.

68. Sode K, Brodelius P, Meussdoerffer F, Mosbach K and Ernst JF. Continuous production of somatomedin C with immobilized transformed yeast cells. Appl Microbiol Biotechnol 1988; 28:215-221.

IMPROVED BIOMASS HOLDUP

This chapter is dedicated to the memory of Dr. Fiona J. Morgan who sadly died during the autumn of 1991 just three and a half years after completing her PhD.

The performance of a fermentation system is dependent on the volumetric rate of reaction, which can be defined as the product of the net specific rate of reaction and the amount of biomass in the vessel, the biomass holdup. For a continuous stirred tank reactor (CSTR) where the cells are freely suspended, this can be readily estimated from a small sample of the effluent. This results from the fact that microbial cells have a similar density to that of the bioreactor broth and are of microscopic size, so that, in a well mixed vessel, any element of the bioreactor broth will contain an equal concentration of cells. For a CSTR the biomass holdup may also be predicted from consideration of the growth kinetics and is usually a function of the dilution rate (D), the maximum specific growth rate (μ_{max}) and the inlet concentration of the limiting substrate. In such a system biomass holdup falls to zero as D approaches μ_{max}. This condition, referred to as washout, imposes a severe limitation on the continuous operation of free cell systems.

In immobilized cell systems the situation is rather different, as indicated by Morgan[1] in a review of biomass holdup in immobilized cell reactors:

"Unlike other bioreactor systems the amount of biomass present in an immobilized cell reactor is an almost independent variable and as such it is a major factor in performance assessment. Immobilized cells are usually in the form of discrete particles so that determination of the biomass holdup is not as straightforward as it is for a conventional suspended cell system."

In the immobilized cell system biomass holdup may be defined as the product of the number of support particles in the reactor and the average biomass holdup per particle, plus any freely suspended cells that are present. Without the restriction imposed by washout, the potential exists with such systems for greatly improved biomass holdup. This chapter deals with some aspects of this improved biomass holdup.

A considerable diversity exists in the size and shape of particles used in immobilized cell reactors, and in the mode in which these reactors are operated. The validity of using an average value for particle biomass holdup to predict the volumetric rate of reaction will therefore depend on the system being described. In general, biomass is either in the form of fixed films on particle surfaces or is distributed throughout the particles. Similarly, the particles themselves are either fixed in place within the reactor or are distributed throughout the reactor. Thus control of biomass holdup can be achieved through control of the number of particles per unit volume within the reactor or through control of the amount of biomass associated with each particle.

PARTICLE HOLDUP IN THE REACTOR

It is generally desirable to minimize bioreactor volume for any given duty. A major factor in achieving this for an immobilized cell system is in the number of support particles that can be accommodated per unit volume (particle number density). By maximizing the particle number density the proportion of the reactor volume not occupied by cells (referred to as the reactor voidage) is kept to a minimum. It is not appropriate, however, to simply fill the reactor volume full of particles and generally some particle movement is desirable. This limits the number of particles per unit volume which can be achieved in practice.

In order to suspend and circulate particles, there has to be sufficient energy input to the liquid to overcome gravitational and buoyancy forces, and to cause movement through the liquid. Thus parameters affecting the magnitude of these forces and also the flow characteristics of the liquid over, or through the particles will be important. These include particle characteristics such as size, density, shape, porosity, surface condition (degree of smoothness), and surface behavior (wettability). They also include reactor configuration parameters which alter, for example, the size and shape of the circulation loop, or the way in which the gas enters the liquid, i.e. aspect ratio, scale or gas distributor design. Similarly, the properties of the liquid within which the particles are suspended will be important, namely density, viscosity and surface tension. In many systems, the only energy input to the system is via sparged gas. This will have a direct effect on the circulatory behavior and therefore the operating range of particle holdup (Example 5: Effect of Reactor Hydrodynamics on Particle Holdup).

In practice, a number of the above parameters are nominally fixed; namely the reactor configuration and the liquid medium being chosen or required for a particular duty. In addition, the practical size range of particles that may be used is limited and is of the order 3-25 mm. The lower end of this range is generally preferred for immobilized cell reactions (in order to minimize diffusion path length), whereas larger sizes are preferred for ease of particle handling. If particles contain significant amounts of biomass then it is inevitable that their density will be similar to that of the cells (and usually the liquid) and that they will therefore be relatively neutrally buoyant. Processes for manufacturing particles generally result in either spheres (droplets) or involve cutting large blocks, first into sheets, then into smaller (cuboidal) particles so that particle shape does not vary too widely.

BIOMASS HOLDUP IN THE PARTICLE

The biomass holdup per particle, on the other hand, though it too may be affected by the mode of operation of the reactor, in particular the hydrodynamic shear, largely depends on the technique used for immobilization.

For systems in which cell growth occurs, the actual amount of biomass associated with an individual particle will ultimately depend on the organism being immobilized and the conditions within the bioreactor. With active immobilization techniques the biomass holdup at reactor start-up is predetermined, while for passive immobilization it is essentially zero. During reactor operation biomass holdup is determined both by biological parameters such as growth rate, and by physical parameters such as hydrodynamic shear. Under certain conditions, diffusional limitations within the immobilized biomass film or floc can lead to sloughing or to 'hollow centers.'

There are, however, many other factors which affect the biomass holdup within an immobilized cell reactor. These factors are analyzed below:

Support Matrix

The volume fraction occupied by the carrier material will effectively dilute the cell concentration in the particle.

Cell Distribution

Nutrient depletion and/or product inhibition may result in a heterogeneous distribution of the biomass with a densely populated zone near the particle surface and a less populated or empty particle interior. This zonation, although beneficial from an effectiveness factor perspective, is inefficient as far as space utilization is concerned. Numerous authors have reported such spatial heterogeneities, especially so for in situ cell cultivation.[2] Data from the group of Robertson[3-5] indicate that there was a thin layer of metabolically active cells adjacent to each fiber in microporous hollow fiber reactors, while cells located in regions more than 30μ from the fibers are metabolically stressed because of nutrient limitation. Moreover, it was observed that the volume expansion by the cells as a consequence of cell proliferation induces convection of cell mass out of the growth region into a region of the reactor filled with starving cells, which then accumulate in the reactor. In immobilized algal systems, the penetration of light is important in determining the size of the growth region, though it has been demonstrated that the supply of carbon dioxide is more important than light intensity.[6]

Electrical Charge

Particle charge has been reported to affect cell adhesion in attached cell systems. For example Michaux et al[7] reported that the yield of cell immobilization (*Saccharomyces*) on sawdust varied from 94.2 to 145.8 mg/g of support with addition of gelatin; this was attributed to the gelatin giving a positive charge to the cells and a negative charge to the carrier. The charge density of microcarriers appears to be important for the attachment and growth of mammalian cells, though its importance may have been overestimated.[8]

Pore Size

The pore size of the particle is important because it determines its 'leakiness' and thus the biomass holdup. Messing and Oppermann[9] demonstrated that there was a maximum in the yield of biomass per mass of carrier and this maximum was dependent on the method of reproduction of the microbes (fission, budding) as well as the pore size of the carrier. In the physical entrapment of cells in biomass support particles (BSPs) the pore size determines the amount of cells that can penetrate into the matrix as well as the depth of the matrix that can be effectively used for entrapment (Example 6: Effect of Pore Size on Biomass Holdup in BSPs). Excessive cell leakage may be prevented by, for example, coating an alginate gel with urethane polymer.[10] It is questionable, however, how long a 'non-leaking' particle with growing cells can be sustained.

Attrition

The rate of attrition in biofilm reactors which depends mainly on hydrodynamic conditions in the reactor, will clearly affect the amount of biomass holdup achievable during steady operation (Example 10: Development of an Immobilized Cell Reactor for Use with Filamentous Fungi, chapter 6).

Interactions Between the Cell and the Support

The biological interactions between some supports and certain types of immobilized cells is important in determining the biomass holdup in the particle. For example, Hayman et al[11] reported that the use of collagen lattices in the immobilization of eukaryotic cells had a positive effect on the survival and growth of hybridoma and fibroblastic cells which reached tissue-like densities. This was attributed to the collagen being a major constituent of extracellular matrix in vivo and, thus, promoting cell adhesion and growth in vitro. Co-immobilization of microorganisms with their growth promoters inside agar has been reported to yield a high cell density.[12] Cell immobilization in κ-carrageenan with tricalcium phosphate increased the biomass concentration, possibly due to advantageous pH effects.[13] An extreme case of ecological advantage is when the carrier is the main energy source for the cell as, for example, in cellulose degradation by *Ruminococcus albus*.[14]

CELL CHARACTERISTICS

Various biological characteristics of the immobilized cells (e.g. type, history, charge, size, shape, method of reproduction, composition of the cell wall, oxygen requirements) are crucial in attaining high biomass concentrations. As an example of the effects of cell history, Figure 3.1 shows the effect of spore inoculum age for *Penicillium chrysogenum* cells on the degree of immobilization achieved in BSPs.[15] The degree of immobilization in this case is defined as the percentage of total biomass in the immobilized state, the remainder being assumed to be in free suspension.

As can be seen, there was a significant influence of spore age on immobilization, for spores older than 8 days. Clearly, after a certain age the spores begin to lose their ability to attach to the particles or to surfaces in general.

ENVIRONMENTAL CONDITIONS

The physico-chemical conditions of the surrounding solution, e.g. ionic strength, temperature and pH, are reported to affect primarily adsorbed cell systems. In addition, hydrodynamic conditions are clearly important. Figure 3.2 shows the effect of rotation

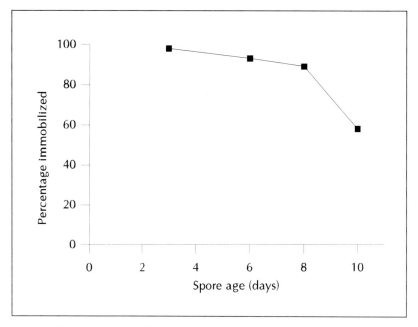

Fig. 3.1. Effect of spore inoculum age on the degree of immobilization of Penicillium chrysogenum *cells in 6 mm polyurethane BSPs (60 ppi) in shaken flasks.*

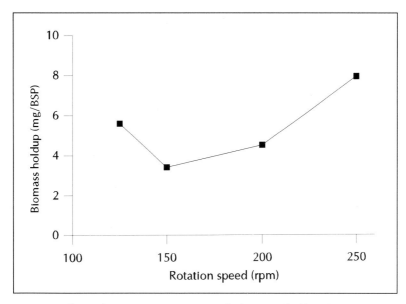

Fig. 3.2. Effect of agitation rate on particle biomass holdup for Penicillium chrysogenum *cells immobilized in 6 mm polyurethane BSPs (60 ppi) in shaken flasks.*

speed on biomass holdup for immobilized cells of *Penicillium chrysogenum* in shake flasks.[15]

The results are interesting in that both the lowest and highest rotation speeds gave high immobilized cell concentrations. Better oxygen and other substrate mass transfer, although being a factor that could have influenced the results obtained at 250 rpm, would not explain the results at 115 rpm. Also, the fact that the suspended cell concentration was lower at 250 rpm than at 150 or 200 rpm, suggests that another factor may have been responsible for the large immobilized cell concentration obtained.

The effect of different hydrodynamic conditions in the flasks at different rotation speeds seems to have been the main reason for the results obtained. Although the largest rotation speed was double the lowest, the particle movement was essentially identical in both cases. At 115 rpm, the liquid was swirling around the flask, but the particles were not exactly being pushed by it. Instead, they were being 'bathed' by the liquid, their position with respect to the flask was not changing significantly. At 250 rpm, the liquid was swirling as a bulk around the flask carrying the particles with it, as if they were being centrifuged. This time, the

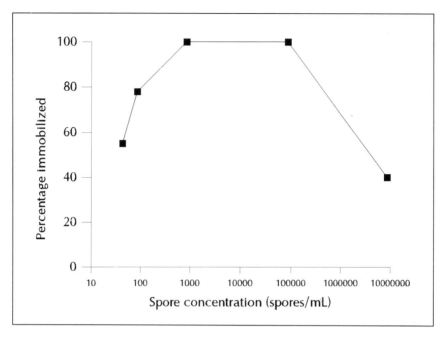

Fig. 3.3. Effect of inoculum spore concentration on the degree of immobilization of Penicillium chrysogenum *cells in 6 mm polyurethane BSPs (60 ppi) in shaken flasks.*

position of the particles with respect to the liquid was not changing. At other rotation speeds, the particles were seen to tumble around the flask. It can therefore be considered that at the lowest and highest rotation speeds, the turbulence caused by agitation was not high and that these gentle conditions encouraged immobilization.

Initial Conditions

Immobilized cells are dynamic systems and as such depend heavily on the initial conditions they are subjected to. A typical example is the initial biomass concentration, or in the case of sporulating organisms, the initial spore concentration in the particle. The effect of the inoculum spore concentration on the degree of immobilization achieved in 60 ppi polyurethane foam BSPs is shown in Figure 3.3.[15]

It can be seen that below a spore concentration of about 10^3 spores/mL increasing spore concentration resulted in increasing degree of immobilization, in terms of the percentage of total biomass immobilized. Immobilization of all the spores occurred at this concentration but above a spore concentration of around 10^5 spores/mL, the percentage of immobilization decreased again as more spores remained in suspension. This increased suspended spore concentration could indicate a saturation level for the BSPs or might be more closely linked to the tendency of spores to aggregate prior to and during immobilization.

It is likely that successful immobilization depends on aggregates of sufficient size becoming entrapped within the pores of the BSPs. Such aggregation is a random process based on interparticle collisions and will therefore be concentration dependent. At low concentrations, aggregates may not become large enough for entrapment to be assured and hence the degree of immobilization is less than 100%. On the other hand at very high spore concentrations, aggregation could result in some particles becoming too large to become entrapped within the BSP pores. Hence there is an optimum inoculum size for full immobilization.

Clearly the initial conditions, in terms of inoculum size, have some influence on the degree of immobilization achieved and therefore ultimately the biomass holdup.

HIGH BIOMASS CONCENTRATION

In a number of reports the biomass concentration in immobilized cell bioreactors has been shown to be substantially higher than the biomass concentration in conventional free cell bioreactors.

On a fairly large scale, the concentration of *S. cerevisiae* cells in Ca-alginate beads employed for pilot-plant ethanol fermentations was estimated as more than 250 g dry cell/L gel.[16] Similar concentrations have been reported by other investigators: 300 g/L for yeast immobilized in Ca-alginate.[4] Given that the gel particles will occupy approximately 50% of the reactor volume, this is equivalent to around 150 g/L throughout the bioreactor. High yeast concentrations for a small scale brewing system in which the cells were immobilized into stainless steel BSPs using a filtration technique can be achieved (Example 7: Colonization of BSPs by Filtration).

The maximum concentration of mouse hybridoma cells immobilized in Ca-alginate beads in an airlift reactor reached a value of 2×10^8 cells/mL capsule.[18]

The relatively high biomass concentrations within immobilized cell particles may lead to enhanced reaction rates and, thus, reduced reactor volumes.[19] A combination of high cell concentration in conjunction with continuous operation at high dilution rates may also reduce the risk of microbial contamination. The use of immobilization reduces or eliminates the fear of losing the microbial population by culture-washout which can be caused by, for example, operating a CSTR at high dilution rates or by introducing an inhibitory compound in the reactor. High biomass concentrations can, however, also be achieved by cell recycling, i.e. by separating the cells from the effluent and recycling them back to the bioreactor.

An interesting observation is that immobilized cells may distort their shape and/or reduce their volume and this may lead to an increase in the packing- or dry-cell-density above the limit which is often assumed for freely-suspended cells. This can be attributed to the pressure exerted from other cells and the immobilization matrix. It has been demonstrated[20] that growing entrapped *E. coli* can exert a pressure of at least 3 atm on their surroundings which results in reduced cell volume leading to a dry cell concentration of 430 g/L; cell densities as high as 850 g/L were measured when aggregates of starved *E. coli* were subjected to controlled applied stresses of 9 atm.

Very high cell concentrations have been reported with hollow fiber reactors. Chung et al[21] reported a concentration for *Nocardia mediterranei* immobilized in a dual hollow fiber bioreactor of 550 g/L. The local cell concentration of *E. coli* K-12 cells confined in microporous hollow fiber membranes was reported to be greater than 400 g/L, in excess of the predicted limit based on the specific volume of free cells determined by tracer exclusion.[3] The concentration of plant cells (*Lithospermum erythrorhizon*) in a dual hollow fiber bioreactor reached 325 g/L.[22] Dhulster et al[23] studied the growth of a plasmid-harboring strain of *E. coli* immobilized in κ-carrageenan and observed a very high cell density within the cavities of the gel (1.7×10^{11} cells/mL) at least two orders of magnitude higher than the cell density in suspension (8×10^{8} cells/mL).

Forced substrate supply has been used in order to increase further the biomass concentrations in immobilized cell particles. Yeast cells immobilized in a gel layer with forced substrate supply exhibited an heterogeneous cell distribution with a maximum of

1.7 x 10^9 cells/mL at the surface of the gel layer.[24,25] Assuming that 10^9 yeast cells correspond to a dry cell weight of 25 mg, the calculated dry cell density is 425 g/L. The average cell density in a 14 mm gel layer with forced substrate supply was estimated to be 200 g/L, four times higher than normal immobilized cell systems. The cellular distribution in the gel layer was analyzed by model equations taking into account the inhibitory and toxic effects of ethanol on the growth of yeast cells.

EXAMPLE 5:
EFFECT OF REACTOR HYDRODYNAMICS
ON PARTICLE HOLDUP

The total amount of biomass in an immobilized cell bioreactor depends not only on the biomass holdup per particle but also on the number of particles in the reactor. Maximizing the latter is a challenging job, especially for non-conventional bioreactor configurations. In this study, a circulating bed reactor (CBR) containing a range of near neutral buoyancy particles was studied at a range of aspect ratios and superficial gas velocities in order to identify the reactor hydrodynamic conditions leading to maximum particle holdup in the reactor. The study involved varying superficial gas velocities from 0 - 0.015 m/s at aspects ratios between one and three. Semi quantitative data were obtained for the minimum superficial gas velocity required to circulate particles in several systems, with either fixed particle type and varied aspect ratio, or fixed aspect ratio and varied particle type.

MATERIALS AND METHODS

The CBR test rig consisted of a rectangular cross-section perspex tank with a gas chamber at its base. The tank, constructed in three sections from 10 mm thick perspex sheet, is illustrated in Example Figure 5.1. The lower (gas chamber) and middle sections, separated by the aeration plate, were held together by bolted flanges. The upper section of the tank was positioned and held in place with silicone rubber sealant when aspect ratios above 1.5 were required. The aeration plate was 3 mm thick perspex sheet through which 35 x 0.5 mm orifices were drilled. The number of orifices was chosen according to the procedure outlined by J.F. Dean (PhD Thesis, UMIST, 1991).

A variety of particle types were used in the studies. The spherical particles were used as they were "ideal" particles, being smooth, solid spheres of known density. Empty BSPs (rough, porous cuboids) were used as this would be the condition they would be in at the beginning of a fermentation, and BSPs filled with agar (rough, solid, cuboids) to simulate immobilized biomass were used. The empty BSPs were boiled for several hours to reduce the hydrophobic characteristics they exhibit when new. They were subsequently stored under water. BSPs were filled with agar gel (approximately 2% w/v) by autoclaving in a solution of agar, squeezing out any entrapped air while submerged then, as the agar cooled and was about to set, removing the BSPs and spreading them on a cold surface to harden. They were then stored under water to prevent them from drying out.

The tank was filled with the required number of particles and volume of tap water to achieve the desired solid holdup and aspect ratio. The air flowrate was increased until all the particles were circulating readily. The system was allowed to stabilize for several minutes. The air flowrate was then gradually reduced in small increments. After each increment the system was allowed to stabilize and the particles were observed to determine whether any of them had ceased to be mobile. If no particles had become stationary a further reduction was made to the air flowrate. If some particles had become stationary, then the air flowrate was noted and increased until all particles were moving again. The procedure was then repeated, using smaller increments. This process was repeated, as necessary, until the air flowrate at which the particles just began to remain stationary was determined. Particles with positive buoyancy became stationary at the top of the tank and were judged to be stationary if they stayed at the top and were never entrained into the downflow. Such "stationary" particles were not, however, necessarily

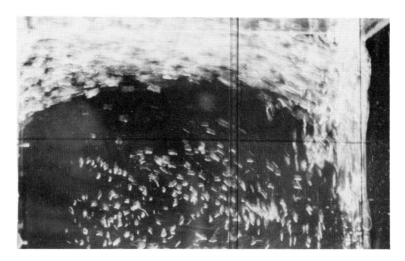

Example Fig. 5.1. The circulating bed reactor.

totally motionless. A similar criterion was applied to the particles with negative buoyancy, which became stationary at the bottom of the tank. It was slightly more difficult to determine if a particle was stationary at the bottom of the tank because of the "bumping" caused by bubbles as they emerged from the orifices. The particles were considered stationary if they never entered the rising flow.

RESULTS AND DISCUSSION

Effect of aspect ratio

Example Figure 5.2 shows the maximum circulating particle holdup plotted against superficial gas velocity for a range of aspect ratios (height/diameter). It can be seen from Example Figure 5.2 that increasing the aspect ratio from a half to unity causes a dramatic decrease in the superficial gas velocity required to cause particles to circulate. Above an aspect ratio of unity there is only a minor decrease in the gas velocity with aspect ratio.

Effect of particle type

Example Figure 5.3 shows the maximum circulating particle holdup against superficial gas velocity for a range of neutral buoyancy particles. It can be seen that all the particles produced similar *S*-shaped curves. The

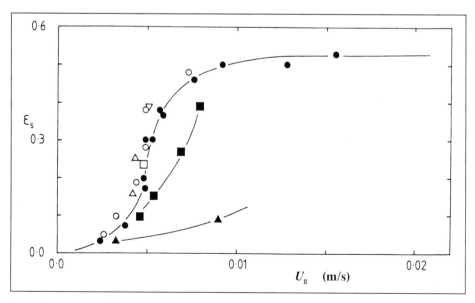

Example Fig. 5.2. Solids holdup, ε_s' against superficial gas velocity, U_g(m/s) required to circulate the particles over a range of aspect ratios for 15.5 mm polyethylene spheres. Aspect ratios: (▲) 0.5; (■) 0.75; (●) 1.0; (O) 1.5; (□) 2.0; (△)3.0; (▽) 3.7.

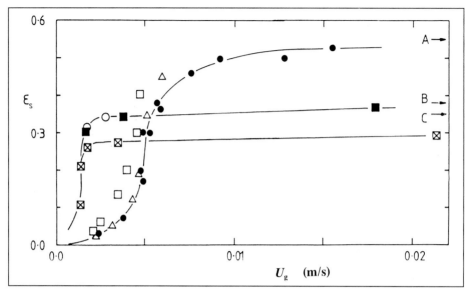

Example Fig. 5.3. Solids holdup, ε_s' against superficial gas velocity, U_g (m/s) required to circulate the particles at an aspect ratio of unity with a range of particle types; (□) 13.0 mm polyethylene spheres; (●) 15.5 mm polyethylene spheres; (△) 18.4 mm polyethylene spheres; (⊠) 10 mm cubic BSPs empty; (■) 10 mm cubic BSPs, agar filled; (O) data from operating CBRs. Lines A, B and C represent packed bed solids holdup for polyethylene spheres, agar-filled BSPs and empty BSPs respectively.

maximum circulating number density approached that of a packed bed. With reticulated foam BSPs the packed bed number density depended on whether the particles were empty or filled with agar gels (to simulate immobilized biomass). Far lower superficial gas velocities were required to circulate BSPs compared with polyethylene spheres. This may be due to their differing terminal velocities. As the BSPs have far lower terminal velocities, far less energy (from the sparged gas) could be expected to be required to achieve circulation.

CONCLUSION

The particle holdup in a CBR can approach that of the packed bed reactor, albeit at higher superficial gas velocities and therefore at higher energy costs. A CBR seems to provide a good compromise between biological activity and process stability.

ACKNOWLEDGMENT

The material for this example was adapted from the PhD Thesis of Dr. Jonathan F. Dean (UMIST, 1991).

EXAMPLE 6:
EFFECT OF PORE SIZE ON BIOMASS HOLDUP IN BSPs

Many studies of the effects of pore size on biomass holdup in BSPs have been carried out in our laboratory; results for three cell types are presented in this example.

BACTERIA

The model bacterial system used to investigate the interactions between the particle structure and biomass holdup comprised cells of *Streptomyces longisporoflavus* immobilized in foam BSPs (FJ Morgan, PhD Thesis, UMIST, 1989). Foam BSPs are available only at fixed pore sizes ranging from 10-100 ppi (nominal pore size = 254 - 2540 mm). The results presented here show the effect of pore size on the growth of *S. longisporoflavus* in shake flasks and reactors.

The results from the shake flask experiments are presented as average values for three replicate flasks. In the first experiment, shake flasks containing foams of one pore size only, were inoculated, incubated for 162 h and then analyzed. Example Figure 6.1 shows the change in immobilized cell concentration as the pore size increased. It emerges that biomass holdup attains a maximum in the range between 250 - 600 mm; above 600 mm there appears to be a weak inverse

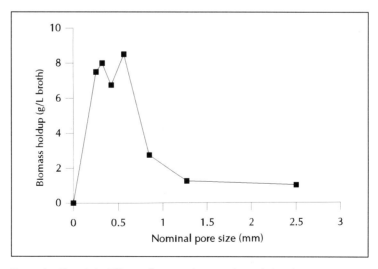

Example Fig. 6.1. Effect of pore size on immobilized Streptomyces longisporoflavus *concentration in shake flasks. Nominal pore sizes: 100 ppi (0.254 mm); 80 ppi (0.318 mm); 60 ppi (0.423 mm); 45 ppi (0.564 mm); 30 ppi (0.847 mm); 20 ppi (1.27 mm); 10 ppi (2.54 mm).*

relationship between foam pore size and immobilized cell concentration. The biomass concentrations reported in Example Figure 6.1 were obtained 162 h after inoculation. Prior to this, however, four distinct stages in the immobilization process could be observed:

1. *0 - 20 h: Physical entrapment.* With the exception of 10 ppi foams, biomass became rapidly entrapped within the particles.
2. *20 - 72 h: Cell growth.* The immobilized cell concentration in all the foams increased. However, in the 80 and 100 ppi foams cell growth was preceded by a reduction in immobilized biomass. This may have been due to the washout of some mycelial flocs before they became fully attached by entanglement and growth. The biomass concentration in the 80 and 100 ppi foams continued to increase during this stage.
3. *72 - 116 h: Mycelial fragmentation.* During this stage, all the foams shed substantial amounts of their immobilized biomass. This was probably due to mycelial fragmentation which usually occurs during this period. This loss occurred earlier in the foams with the larger pore sizes.
4. *116 - 162 h: Recolonization.* In the last stage, some recolonization of the foams by the organism took place. This may have been due to growth of the biomass already immobilized in the foams or to fresh uptake of newly grown mycelial flocs.

The changes in immobilized biomass concentration during fermentations in the CBR were more difficult to measure. Analysis of the pore size of the particles sampled showed that the number removed by sampling was inversely proportional to pore size, thus foams containing less biomass were more likely to be sampled. The method of sampling relied on the upward movement of the particles with the culture broth when the reactor was pressurized; the foams containing more biomass were more dense and so required a greater fluid velocity to raise them than the more buoyant empty particles. This sampling method was not therefore providing a representative sample of particles. To obtain a more accurate picture of the immobilization and growth of immobilized biomass in a CBR, several fermentations were stopped at different time intervals and all the particles used for subsequent analysis. This method was not ideal because of the large variation in biomass concentration which can occur between fermentations. However, duplicate fermentations inoculated with the same inoculum culture were carried out for each time interval. The immobilized cell concentrations were much lower in the reactors than those obtained in shake flasks. Although the pattern of immobilization initially appeared to be similar, with rapid physical entrapment in the first 20 h, there subsequently appeared to be no growth and no effect of mycelial fragmentation. The inverse relationship between foam pore size and immobilized cell concentration is clearly shown in Example Figure 6.2 which illustrates the final immobilized cell concentrations for several fermentations in the CBR.

These results support the suggestion that the amount of biomass which becomes immobilized in foam particles is dependent on the degree of protection given from hydrodynamic shear forces. An inverse relationship between foam pore size and immobilized cell concentration was found throughout fermentations in both shake flasks and CBRs, indicating that the increased mesh in the smaller pore sizes offers increased protection from hydrodynamic shear forces. In shake flasks, where the hydrodynamic shear forces are probably lower and less unpredictable than those found in the CBR, much larger amounts of biomass became immobilized, the biomass filled the particles to the extreme edges and even overgrew them. In the CBR where the hydrodynamic shear conditions are increased, much less biomass became immobilized, probably because of the increased shear effects and also because of abrasion of the particles. This could also explain the

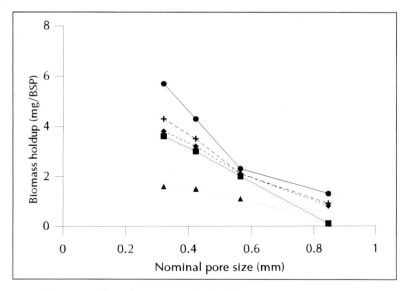

Example Fig. 6.2. Effect of pore size on the final immobilized cell concentration for Streptomyces longisporoflavus *fermentation in a CBR. Symbols refer to five separate fermentations.*

apparent lack of growth of immobilized biomass in the CBR; if initially the immobilized biomass was located towards the center of the particle, growth would fill the edges of the particle but the effects of shear and abrasion would cause attrition of this biomass giving, in effect, a constant immobilized cell concentration. The lack of effect of mycelial fragmentation in immobilized cell fermentations in the CBR can also be explained in these terms. If most of the biomass was immobilized in the center of the particle it would be more difficult for this biomass to be washed out when mycelial fragmentation occurred, whereas in shake flasks loss of biomass immobilized towards the edges of the particle would occur and thus the effect of mycelial fragmentation would be more noticeable. The loss of biomass from the foams with larger pore sizes in between the periods of physical entrapment and growth and also at an earlier stage during mycelial fragmentation also suggest that it is easier to remove immobilized biomass from foams with large pore sizes, i.e. that they offer less protection against the effects of hydrodynamic shear.

Yeast

Three different strains have been used to investigate the effect of particle pore size on biomass holdup for yeast cells; *Saccharomyces cerevisiae* NCYC 1119, *S. uvarum* NCYC 341 and *S. uvarum* ATCC 26602. Experiments with immobilized cells have been carried out both in shake flasks and in a CBR.

All work was carried out using a carbon-limited medium with glucose as the main carbon source and yeast extract as a nitrogen source. The procedure for determining the biomass content (dry weight basis) of foam BSPs involved taking ten particles which were then squeezed with tweezers and washed in distilled water until all cells had been removed. The resultant cell suspension was filtered through a dried and weighed GFA filter paper which was then dried again at 90°C for 24 h and reweighed.

Shake flask experiments were carried out in 500 mL Erlenmeyer flasks, each containing 40 BSPs (ten stainless steel BPSs; ten 30 ppi foam BSPs; ten 45 ppi foam BSPs: ten 60 ppi foam BSPs) and 150 mL medium (glucose concentration of 100 g/L). Each flask was inoculated with 5 mL suspended cells washed from an agar slope and incubated at 30°C for 48 h on a rotary shaker at 150 rpm. At the end of this period, the amounts of immobilized and freely suspended biomass were measured along with the concentrations of glucose and ethanol in the supernatant broth.

In the CBR, the pore size of the foam BSPs was found to be an important factor in the accumulation of cells within the particles. Under all conditions, the 60 ppi particles contained more biomass than the 45 ppi particles which themselves contained more than the 30 ppi particles. Example Figure 6.3 shows a set of data, obtained under steady state bioreactor conditions, in support of this observation. The highest value of *particle biomass holdup, m_p,* recorded for foam particles during continuous operation was 32.5 mg.

Example Table 6.1 shows that for the three yeast strains listed, considerable amounts of biomass were accumulated within the BSPs; the other yeasts tested gave poor results in this respect. While it is not yet clearly understood which characteristics of yeast are associated with their ability to occupy BSPs, it is evidently not due solely to the provision of a quiescent environment, otherwise all the yeasts would have given similar results. Flocculent and adhesive characteristics are likely to be contributory factors, though flocculence cannot be the sole determining factor since most of the original fourteen strains were considered to be flocculent. It is unusual for yeasts to exhibit adhesive qualities and so this too is unlikely to be the major factor. The type of matrix material, stainless steel or polyester foam, appears to have some effect on the extent of cell immobilization. Example Figure 6.3 also shows that the biomass holdup in the CBR is affected by pore size. There is a general

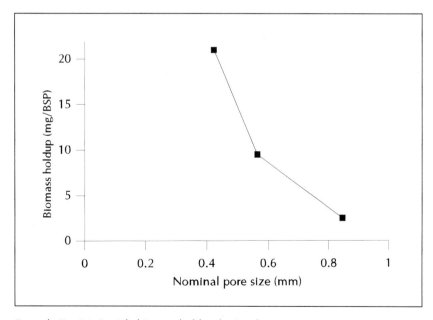

Example Fig. 6.3. Particle biomass holdup for Saccharomyces cerevisiae *NCYC 1119 in foam BSPs, obtained under steady state conditions in a CBR.*

Example Table 6.1. Effect of pore size (ppi) on biomass holdup (g/L) for BSPs from shake flask cultures of three yeast strains

	Pore size (ppi)		
Yeast strain	**30**	**45**	**60**
S. cerevisiae NCYC 1119	31	67	105
S. uvarum NCYC 341	5	14	43
S. uvarum ATCC 26602	19	14	23

trend of increasing holdup with decreasing pore size and this suggests that even smaller pore sizes may be advantageous. The same trend is shown by the shake flask results in Example Table 6.2.

FILAMENTOUS FUNGI

The model fungal system used in this work comprised penicillin producing cells of *Penicillium chrysogenum* immobilized in foam BSPs of pore sizes 30, 45, 60, and 80 ppi (M.J.S. Kelly, MSc Dissertation, UMIST, 1988).

Flasks containing 200 mL of defined medium (Lactose 10 g/L) and 200 BSPs, and controls containing no BSPs were inoculated with spores and grown at 25°C on a rotary shaker. After 24 hours, growth appeared as micro-colonies in the control and to a lesser extent in the flasks containing BSPs. Visible growth occurred in the matrix of the 30 and 45 ppi BSPs after 48 hours. Presumably there was also growth in the 60 and 80 ppi BSPs at this stage, but this was invisible due to the more closely knit structure of these particles. Visible yellow pigment (chrysogenin) appeared after 72 hours. At 240 h no free cells were visible in the immobilized cell flasks.

The effect of pore size was studied in shake flask experiments by comparing growth and penicillin production. Experiments were conducted in duplicate; flasks contained 200 ml of defined medium and 200 BSPs. Control flasks contained 200 ml of defined medium and no BSPs. 30, 45, 60, and 80 ppi BSPs were used. Samples were taken and pH, lactose, penicillin and free cell concentrations and BSP cell masses were determined at intervals of 24 hours for a period of 192 hours. Changes in concentrations and cell mass with time for controls and flasks containing BSPs of different pore sizes are shown in Example Figure 6.4. Growth in the control cultures was in the form of homogenous mycelial growth whereas in the presence of BSPs, free cells grew in the form of dense pellets. These pellets reached 1-2 mm in diameter by the end of the fermentation. The final concentrations of free cells in the flasks containing BSPs was found to be in the region of 2 g/L: less than half that of the controls (Example Fig. 6.4). BSP cell masses increased with decreasing pore size to a maximum of 6 g/L in the 60 ppi flasks. The 80 ppi BSPs showed reduced cell masses.

CONCLUSION

When using inert porous supports for immobilization, such as BSPs, pore size is clearly an important parameter affecting the level of immobilization achieved and therefore the biomass holdup. As such, pore size should at least be quoted in reports and publications but should, preferably, be optimized for each system.

ACKNOWLEDGMENTS

Some of the material for this example was adapted from the PhD Thesis of Dr. Fiona J. Morgan (UMIST, 1988) and the MSc Dissertation of Mr. Mark J.S. Kelly (UMIST, 1988).

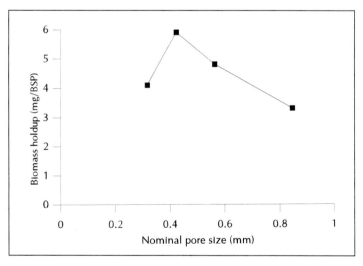

Example Fig. 6.4. Effect of pore size on biomass holdup for Penicillium chrysogenum *cells immobilized in BSPs after 192 hours in shake flasks.*

EXAMPLE 7:
COLONIZATION OF BSPS BY FILTRATION

It is generally acknowledged that high yeast concentration is the most efficient way to achieve rapid fermentation. The feasibility of beer production using BSPs largely depends on how much yeast can be retained in the BSPs during the fermentation period. Biomass holdup in each BSP is determined by the particle volume, particle porosity and cell density. For the immobilization of yeast within stainless steel BSPs, the absolute maximum value of cell density is that of pressed yeast cake, i.e. around 300 g/L and thus the maximum value of biomass holdup would be 27 mg for a 0.113 mL particle with 80% porosity.

During the course of the fermentation, biomass holdup is determined from an integral balance between initial immobilization, growth and wash-out. The amount of initial immobilization is a function of the colonization process, which, ideally, should be simple, and carried out in such a way that a high biomass holdup can be achieved in a very short period. The majority of BSP systems rely on cell growth for colonization (Example 1: Biomass Support Particles, chapter 1). In order to remove the dependence of immobilization on growth, yeast slurries were used for direct colonization by filtration in this example.

MATERIALS AND METHODS

Yeast cells in the form of a slurry, pressed cake or dried cells were used in this study. These can be classified as: brewery yeast (118F), pressed yeast (I), pressed yeast (II), dried baking yeast and dried brewing yeast. The slurries of brewery yeast (*Saccharomyces cerevisiae* 118F) were collected from the bottom of cylindro-conical bioreactors at Samuel Webster and Wilsons Brewery (Halifax, UK). Pressed yeast (I) was purchased directly from Boyds Creameries Ltd. (Old Trafford, Manchester, UK) and had been produced for baking. Pressed yeast (II) was provided by J.C. Seed Ltd. (Harrogate, North Yorkshire, UK). It was produced by filtering creamed yeast, collected from selected breweries in Northern England. Both pressed yeasts contained about 30% solid content. Dried baking yeast (Allinson Co., Surrey, UK) was available at health food shops and dried brewing yeast was obtained from Boots, the chemist or a home brew center.

Prior to use for colonization, pressed yeast or dried yeast was suspended in beer or sterilized drinking water to produce a slurry. All yeast slurries and pressed yeasts were kept in a modified freezer at 0°C and checked for viability and contamination at regular intervals.

The colonization test rig comprised a cylindrical column of 200 mm height and 28 mm internal diameter made of Pyrex glass. The bottom end of the column was connected to a quick fit glass connector (QVF) of 40 mm height and 28 mm internal diameter. Approximately 6000 stainless steel BSPs (6 mm diameter) were placed into the column. In order to retain the BSPs, a perforated stainless steel plate was placed between the column and the QVF section. The volume of the whole column was about 77 mL.

A series of colonization trials were performed as follows: The appropriate yeast slurry was removed from the freezer into a glass vessel and kept uniformly suspended by a magnetic stirrer. This slurry was then pumped upwards through the BSP-packed column at a rate equivalent to a superficial velocity of 120 cm/h. The cell concentration of both inlet and outlet flows were monitored during the course of the test to determine the amount of yeast retained within the column. At the end of each test the BSP biomass content was determined.

RESULTS

Typical results for colonization trials with each of the yeast types are shown in Example Table 7.1. It can be seen from these results that considerable increase in cell concentration is possible using the filtration colonization technique. Compared to usual levels of biomass holdup in conventional free cell systems (typically much less than 30 g/L) the BSP immobilized yeast bioreactor retains significantly more biomass.

The key stages in the colonization process can be described as follows: The yeast slurry is passed continuously upwards through a relatively deep column of BSPs initially free of yeast cells. The lowermost layer of BSPs, when contacted with the concentrated slurry, at first retain yeast cells rapidly and effectively. Yeast cells remaining in suspension

Example Table 7.1. Biomass holdups achieved via filtration of various yeast slurries through a packed bed of stainless steel BSPs

Yeast	BY118F	PY(I)	PY(II)	DBY
Viability (%)	88	85	85	90
Slurry Concentration (g/L)	51	72	60	75
Biomass holdup achieved (g/L)	99	97.4	93.8	104

are then substantially removed by the layers of BSPs in the upper part of the column. The effluent from the top of the column is at this stage, practically cell-free. As the slurry continues to flow, the yeast saturation zone expands from the lower to the upper part of the column until most the whole bed is saturated with yeast cells. From this point the system quickly reaches the equivalent of an adsorption column breakpoint, beyond which the yeast concentration in the effluent rises rapidly, since the column is effectively saturated. At this stage, cell entrapment has ceased and the bed is assumed to be entirely in equilibrium with the feed suspension.

CONCLUSION

By colonizing BSPs with yeast cells from slurries, biomass holdups more than 10 times those normally associated with free cell systems can be achieved.

ACKNOWLEDGMENT

The material for this example was adapted from the PhD thesis of Dr. Fan-Chiang Yang (UMIST, 1989).

REFERENCES

1. Morgan F. Optimization of secondary metabolite production by an immobilized *streptomyces* sp. PhD Thesis, UMIST 1988.
2. Wada M, Kato J and Chibata I. Continuous production of ethanol using immobilized growing yeast cells. Eur J Appl Microbiol Biotechnol 1980; 10:275-287.
3. Karel SF and Robertson CR. Cell mass synthesis and degradation by immobilized *Escherichia coli*. Biotechnol Bioeng 1989; 34:337-356.
4. Karel SF and Robertson CR. Autoradiographic determination of mass-transfer limitations in immobilized cell reactors. Biotechnol Bioeng 1989; 34:320-336.
5. Robertson C. Membrane constrained immobilized living cell systems. Polym Prepr Am Chem Soc Div Polym Chem 1986; 27:419-420.
6. Robinson PK, Goulding KH, Mak AL and Trevan MD. Factors affecting the growth characteristics of alginate-entrapped *Chlorella*. Enzyme Microb Technol 1986; 8:729-733.
7. Michaux M, Paquot M, Baijot B and Thonart P. Continuous fermentation: improvement of cell immobilization by zeta potential measurement. Biotechnol Bioeng Symp 1982; 12:475-484.
8. Himes VB and Hu WS. Attachment and growth of mammalian cells on microcarriers with different ion exchange capacities. Biotechnol Bioeng 1987; 29:1155-1163.
9. Messing RA and Oppermann RA. High-surface low-volume yeast biomass composite". US 4149937 1979.
10. Yoshida H, Kamihira M, Iijima S and Kobayashi T. Continuous production of anti-erythropoietin antibody by immobilized hybridoma cells. J Chem Eng Jpn 1989; 22:282-286.
11. Hayman EG, Ray NG and Runstadler PW Jr. Production of biomolecules by cells cultured in tri-dimensional collagen microspheres. Moody GW, Baker PB (eds). Bioreact Biotransform. London: Elsevier Appl Sci, 1987:132-140.
12. Terao Y, Hakubun E and Shioyama M. Immobilization of high-density microorganisms". Jpn Kokai Tokkyo Koho JP 62/32883, 1987.
13. Wang HY and Hettwer DJ. Cell immobilization in κ-carrageenan with tricalcium phosphate. Biotechnol Bioeng 1982; 24:1827-1838.
14. Ohmiya K, Nokura K and Shimizu S. Enhancement of cellulose degradation by *Ruminococcus albus* at high cellulose concentration. J Ferment Technol 1983; 61:25-30.
15. Rodrigues PAL. Feasibility of using immobilized cells in stirred tank fermenters. PhD Thesis, UMIST 1995.
16. Nagashima M, Azuma M, Noguchi S, Inuzuka K and Samejima H. Continuous ethanol fermentation using immobilized yeast cells. Biotechnol Bioeng 1984; 26:992-997.

17. Shao G, Zhao Y, Wang B, Mao H, Zhuo Z and Chen T. Preparation of immobilized yeast cells and its application. Shanghai Keji Daxue Xuebao 1985:76-81.
18. Bugarski B, King GA, Jovanovic G, Daugulis AJ and Goosen MFA. Performance of an external loop air-lift bioreactor for the production of monoclonal antibodies by immobilized hybridoma cells. Appl Microbiol Biotechnol 1989; 30:264-269.
19. Luong JHT and Tseng MC. Process and technoeconomics of ethanol production by immobilized cells. Appl Microbiol Biotechnol 1984; 19:207-216.
20. Stewart PS and Robertson CR. Microbial growth in a fixed volume: studies with entrapped *Escherichia coli*. Appl Microbiol Biotechnol 1989; 30:34-40.
21. Chung BH, Chang HN and Kim IH. Rifamycin B production by *Nocardia mediterranei* immobilized in a dual hollow fiber bioreactor. Enzyme Microb Technol 1987; 9:345-349.
22. Kim DJ, Chang HN and Liu JR. Plant cell immobilization in a dual hollow fiber bioreactor. Biotechnol Tech 1989; 3:139-144.
23. Dhulster P, Barbotin JN and Thomas D. Culture and bioconversion use of plasmid-harboring strain of immobilized *E. coli*. Appl Microbiol Biotechnol 1984; 20:87-93.
24. Mitani Y, Nishizawa Y and Nagai S. Ethanol production by immobilized cells with forced substrate supply. J Ferment Technol 1984; 62:249-253.
25. Mitani Y, Nishizawa Y and Nagai S. Growth characteristics of immobilized yeast cells in continuous ethanol fermentation with forced substrate supply. J Ferment Technol 1984; 62:401-406.

CHAPTER 4

IMPROVED MASS TRANSFER

Normally we associate mass transfer in immobilized cell systems with what goes on inside the particle. This is generally regarded as a negative feature since diffusion limitations normally result in reduced reaction rates. This does not have to be the case and, indeed, reaction rates can be enhanced.[1] For example, the existence of gradients inside an immobilization aggregate may be beneficial for substrate-inhibited reactions. The reason for this is that as the substrate concentration falls with distance within the aggregate, the effect of substrate inhibition becomes locally less significant. This effect is expected to be pronounced for fairly large substrate concentrations at the particle surface and large particles. Such intra-particle mass transfer effects are dealt with in later chapters.

The basis for this chapter is mass transfer outside the particle (i.e. in the liquid phase) which is often better in immobilized cell systems due to the particulate nature of the biomass. As Thomas[2] points out for filamentous fungal systems:

"...the dispersed form of growth can lead to hyphal entanglement in the fermentation broth. This may have an adverse effect on the rheological properties of the broth, making it highly viscous and pseudoplastic, and thus difficult to mix in a large fermenter, so that inhomogeneities might arise. Furthermore oxygen transfer from sparged air bubbles to the micro-organisms decreases with increases in viscosity. In severe cases, this can lead to oxygen limitation. Heat transfer for fermenter cooling can also be reduced under such conditions. These problems can lead to lowered productivity. Such rheological problems do not tend to arise in fermentations with pelleted forms..."

Such pellets are, of course, a form of cell immobilization, albeit a poorly controlled one. When the immobilization is carried intentionally and in a controlled manner the advantages can, in principle, be even greater.

MIXING AND MASS TRANSFER

Good mixing and mass transfer conditions are essential for good bioreactor performance, regardless of the type of fermentation, i.e. free or immobilized. Achieving and maintaining good mixing and mass transfer in the bioreactor is not always a straightforward task; mixing and mass transfer not only are intricately linked with a host of interacting physicochemical and biological parameters but are also heavily dependent on the geometry of the fermentation vessel and its scale of operation. Introducing immobilized cells into the bioreactor is bound to add to the complexity of the biological system and have an effect on the rheological properties of the fermentation broth and, thus, on the efficiency of mixing and mass transfer. In the following sections, the effect on mixing and mass transfer of introducing immobilized cell particles into the bioreactor is analyzed with particular emphasis on gas-liquid mass transfer effects.

Effect on Rheology

The rheological properties of the fermentation broth affect mixing and transport phenomena in the bioreactor and, indirectly, cell physiology and productivity. High concentrations of freely suspended cells, cell morphology and the presence of macromolecular extracellular compounds influence the rheological properties of the fermentation broth. This, in turn, may cause the broth to change its behavior from Newtonian to non-Newtonian, most commonly to shear thinning or pseudoplastic.

There is no shortage of examples in the literature where the rheological properties of the fermentation broth underwent significant changes during the course of a fermentation, especially in fungal fermentations with freely-suspended cells where the microorganism was in its filamentous form. Such changes are often quantified by the consistency index (K) and the power law exponent (n). In a *Streptomyces* fermentation using freely suspended cells, the consistency index K increased 8 times in the first half of the fermentation decreasing two-fold with respect to its peak value in the remainder. In the same fermentation, the power law index n decreased 50% over the whole of the fermentation (ca. 200 hours).[3] A *Penicillium chrysogenum* fermentation broth showed a maximum increase in K of 100-fold whereas n showed a maximum decrease of ca. 50%.[4] One way of controlling the consistency index is by broth dilutions, although this does not lead to significant process improvements.[5] Dilutions with water appear to be more successful in controlling the viscosity.[6]

The key factor affecting broth rheology is the morphology of the cell, especially in fungal fermentations with filamentous micro-organisms. The filamentous form interferes severely with the rheological properties of the broth[7] and, in turn, broths are generally non-Newtonian and become more so as the fermentation proceeds. This makes the system unstable, as the agitation requirements for a suitable mass transfer change during the fermentation. Increasing agitation rates, although improving mass transfer, may cause increasing damage to the hyphae as they go through the high shear impeller region. Changes in rheological properties are less severe if the microorganism is in pellet form, but nutrient distribution inside the pellet is not uniform.[8]

The problem of intra-pellet gradients may be alleviated, at least partly, by immobilizing the fungal cells within a suitable particle of the appropriate size so that the concentration of substrates, nutrients and products within the particle favors the production of the metabolite of interest. Difficult as this may be, it offers significant flexibility in improving the process, much more than trying to control pellet size by, let's say, impeller agitation speed. Cell immobilization clearly has an important role to play in reducing broth viscosity in mycelial fermentations. The reduced viscosity results in decreased power requirements for agitation and aeration and in enhanced gas-liquid mass transfer. Significant oxygen-transfer enhancement accompanied by significant energy savings were reported when mycelial growth was confined to microbeads.[9,10]

Another possible application of immobilized cells is in the production of macromolecular metabolites where the product concentration in fermentations using freely-suspended cells may be limited by the resulting high broth viscosity. Immobilization creates a gradient of product concentration diffusing out of the carrier, and this gradient may be exploited in order to increase the average internal concentration of the product and, thus, its yield. This has been demonstrated by Robinson and Wang[11] who demonstrated that xanthan accumulated inside the immobilization carrier reaching a concentration of 55 g/L -pore volume, while in comparable free-cell fermentations, the maximum achievable density was only 35 g/L.

EFFECT ON OXYGEN MASS TRANSFER

Oxygen mass transfer into the liquid phase is often the limiting factor in bioreactor performance compared to, for example, the rate of mixing.[12-14] The rate of mass transfer into the liquid phase is proportional to the oxygen mass transfer coefficient ($k_L a$) which, in turn, is inversely dependent on viscosity. As mentioned earlier, broth viscosity will often increase during a fermentation, especially in fungal fermentations using freely-suspended cells. As a result, $k_L a$ will decrease.[15,16] The timing for the decrease in $k_L a$ is most unfortunate as it often coincides with increased total oxygen demand brought about by the increase in biomass concentration in the system. This may lead to oxygen deficiency in the bioreactor with damaging effects on production, which can range from a decrease to irreversible loss.[17,18] Some ways of dealing with the problem include oversizing the gas-liquid mass transfer equipment,[13] adding mechanical devices to control foam and increase gas holdup[19] and retro-fitting the impeller; the latter may lead to power savings on the order of 50% for the same $k_L a$.[6]

Cell immobilization offers the potential for improving gas-liquid mass transfer by reducing the viscosity of the fermentation broth. A manifestation of the effect of cell immobilization on oxygen mass transfer can be found in Figure 4.1 where the dissolved oxygen tension and the impeller rotation speed required to maintain an acceptable level of dissolved oxygen in the fermentation broth are plotted for two identical bioreactors containing free and immobilized *Penicillium chrysogenum* cells, respectively. In the free cell fermentation, the rotation speed was increased to 1100 rpm at 40 hours and to 1300 rpm after 54 hours, when the dissolved oxygen concentration was 1.5%. In the immobilized cell fermentation, the rotation speed was increased from 500 rpm to 600 rpm 52 hours after medium replacement, equivalent to a start of fermentation, and stayed at 600 rpm for 38 hours until the end of the fermentation. Note that the increases in rotation speed are

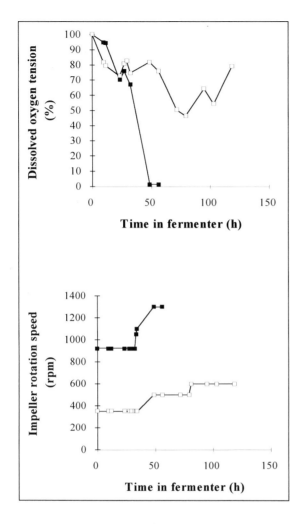

Fig. 4.1. Dissolved oxygen tension and impeller rotation speed in a bioreactor containing only free Penicillium chrysogenum cells (■) and in an identical bioreactor containing immobilized cell particles (□).

equivalent in both cases: 19 and 18% in free cells and 20% in immobilized cells. So, equivalent increases in rotation speed led to effective dissolved oxygen control in the immobilized cell case and had no effect in the free cell case. This highlights not only the improvement in gas-liquid mass transfer but also the better stability and controllability of immobilized cell reactors referred to in chapter 2.

In many immobilized cell systems there is a synergistic effect between freely-suspended and immobilized cells. Improved gas-liquid mass transfer and reduced broth viscosities are not only beneficial for the cells in the immobilized state but also for the freely-suspended ones. For example, the introduction of BSPs in a fermentation medium will reduce the concentration of freely-suspended cells and, thus, broth viscosity. This, in turn, may create favorable environmental conditions for the production of the metabolite of interest by the freely-suspended cells in the bioreactor. An example of the effect of the concentration of freely-suspended cells on their productivity can be found in Figure 4.2; a cursory glance at Figure 4.2 shows that the specific yield of the freely-suspended cells of *S. longisporoflavus* in an intra-cellular antibiotic was much higher at low concentrations of freely-

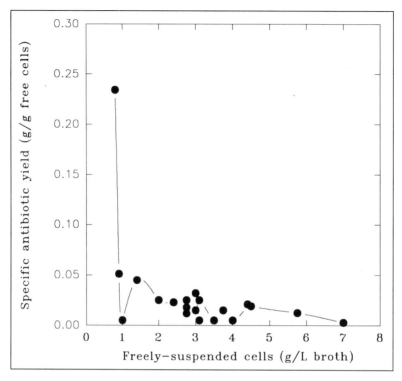

Fig. 4.2. Effect of free cell concentration on the specific yield of an intra-cellular antibiotic by free cells in immobilized S. longisporoflavus cultures in shake flasks.

suspended cells. Being an intra-cellular product, it is very easy to determine the relative contributions of the free and immobilized cells to the overall productivity of the system.

THE PRESENCE OF SOLIDS IN THE BIOREACTOR

So far, it has been argued that the presence of immobilized cell particles in the bioreactor may have a beneficial effect on gas-liquid mass transfer. However, it has to be borne in mind that as the volume occupied by the immobilized cell particles increases (increased solids holdup), the interference with the gas and liquid phases in the bioreactor is bound to increase, resulting in the possibility of lower gas-liquid mass transfer rates. Under such circumstances, a hydrodynamic study of the bioreactor at different values of solids holdup is advisable. An example of such a study for BSP bioreactors can be found in Example 8: Effect of Solids Holdup on Oxygen Mass Transfer.

OTHER EFFECTS

In free cell fermentations, adequate rates of mass transfer may mean high shear stresses which may cause significant damage to cells, depending on the species. Although shear damage can be beneficial in some cases by generating new growth points, it is in most cases responsible for loss or, at least, for considerable decrease in activity.[20-22] This is particularly a problem in stirred bioreactors where the existence of shear gradients means that the effects of shear depend on the time the cells spend in the high shear regions.[21,23] The use of a flow type impeller, like the Prochem Maxflo T, may allow the use of much higher rotation speeds with much less damaging effects of shear.[22] Shear rate can be a critical parameter in scale-up especially for processes dealing with very sensitive cells, like animal cell culture processes.[24] For each culture there is an optimum shear rate that balances eventual damaging effects with suitable operating conditions of mass and heat transfer.[7,23,24] However, there is disagreement on how to quantify this effect and to which shear rate cell damage should be related to. This latter could be the shear rate in the discharge stream of the impeller, the shear rate in the vortices behind the blades or the average shear rate.[25] Of course the use of cell immobilization, and in particular of immobilization techniques where the cells are not in direct contact with the surrounding liquid, protects cells from shear effects and provides an attractive alternative to the cultivation of shear sensitive cells.

In freely-suspended cell systems, micro-mixing problems arise due to the relative size of cells (1 - 5 μm) and of the smallest turbulent eddies (50 - 300 μm).[26] With this length scale, the cells may spend a significant amount of time inside an apparently stagnant eddy where nutrients have been depleted.[26] In this case, the same effects as those caused by the existence of macroscopic concentration gradients can be expected on the cells. The same kind of observations as those occurring in parallel reactions can be made in the case of microorganisms. For example, there may be a larger ratio of the concentration of a product produced under limitation of oxygen to that of a product produced when oxygen is not the limiting nutrient.[27] Cell immobilization significantly increases the relative size of the "cell phase" (i.e. the size of the immobilized cell particles) and is much larger than the size of the smallest turbulent eddies, therefore, no significant micro-mixing problems are anticipated.

Finally, in a free cell system, the density of the cells is similar to that of the surrounding medium which often results in relatively low differential velocities between the cells and the medium and low mass transfer coefficients. Immobilization may increase the apparent density of the cells and, thus, the mass transfer coefficients.

EXAMPLE 8:
EFFECT OF SOLIDS HOLDUP ON OXYGEN MASS TRANSFER

The presence of particles in a bioreactor system is likely to influence gas-liquid mass transfer, particularly if the particles are relatively large and are present in relatively high concentration. This example shows results from studies using BSPs and large buoyant solid particles in both CBRs and STRs.

STUDIES IN GAS AGITATED SYSTEMS

A series of experiments were performed to investigate the effect of solids holdup on oxygen transfer rate in a CBR (J.F. Dean, MSc Thesis, UMIST, 1985). The superficial gas velocity was held constant at 56 cm/min whilst the solids holdup was varied from 0 to 0.60 (using 13 mm spheres) and 0 to 0.30 (using 6 mm BSPs).

The oxygen transfer rate based on reactor volume is shown in Example Figure 8.1 against nominal solids holdup. It can be seen that at low solids holdup values (up to 0.10) the presence of spheres led to an increase in oxygen transfer rate. The behavior of the tank contents appeared very similar to that in the solids free system, with about the same degree of bubble entrainment in the downflow. The particles did not appear to cause the bubbles to coalesce. As the solids holdup was increased the oxygen transfer rate began to fall until it reached a minimum at a solids holdup of approximately 0.30 - 0.45 at which point the particles appeared to have a definite bubble coalescing effect. However, the actual transition from non-coalescing to coalescing was not observed.

Above a solids holdup of approximately 0.45 the spheres began to pack at the top of the tank as the liquid motion was insufficient to cause total circulation. At this stage the oxygen transfer rate was observed to increase again, probably due to increased gas holdup (in the packed cap) and an increased tendency for bubbles to be recirculated.

The presence of BSPs had the effect of gradually decreasing oxygen transfer rate to a similar value as the spheres at a holdup of 0.30 - 0.35, at which point bubble coalescence was again observed.

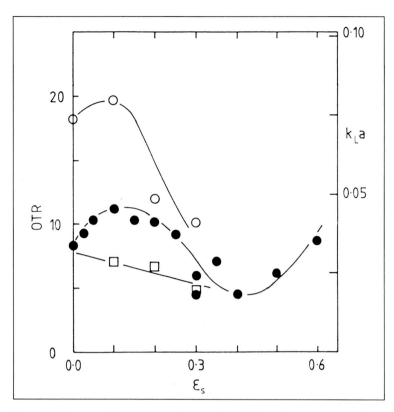

Example Fig. 8.1. Oxygen transfer rate (OTR) based on reactor volume (μmol/L.s) and mass transfer coefficient, $k_L a$ (l/s) against solids holdup, ε_s. (O) 13 mm polyethylene spheres at a superficial gas velocity of 1.62 cm/s; (●) 13 mm polyethylene spheres at a superficial gas velocity of 0.93 cm/s; (□) 10 mm empty cubic polyurethane foam BSPs at a superficial gas velocity of 0.93 cm/s.

STUDIES IN MECHANICALLY AGITATED SYSTEMS

The overall rate of gas-liquid mass transfer is a function of the amount and quality of gas bubbles present in the system, the gas holdup, as well as other parameters such as mass transfer coefficient and concentration driving force. Studies were carried out in STRs to establish the influence of solids holdup on the gas holdup in air-water systems (C. Szabo, MSc Dissertation, UMIST, 1992).

For 15.5 mm polyethylene spheres having a density slightly lower than that of water, Example Figure 8.2 shows a small positive effect of solids holdup on gas holdup over a range of impeller speeds. The particles apparently helped the impeller to disperse the gas phase by breaking up the air bubbles. Visual observations showed that there were much smaller bubbles present in the three phase suspension than in a gas-liquid system without particles, at the same gas flow rate and

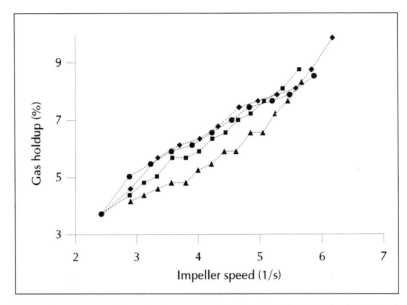

Example Fig. 8.2. Effect of almost neutrally buoyant floating (polyethylene) particles (15.5 mm diameter) on gas holdup for a range of impeller speeds, at a gas flowrate of 0.93 vvm in a 75 L STR. (▲) 0% solids holdup; (■) 5% solids holdup; (◆) 15% solids holdup; (●) 26% solids holdup.

agitation speed. These smaller bubbles can contribute significantly to mass transfer by increasing the interfacial area without necessarily increasing gas holdup.

In studies of the direct effect of solids holdup on mass transfer coefficient, 6 mm polyurethane foam BSPs were used in a 8.5 L STR (P.A.L. Rodrigues, PhD Thesis, UMIST, 1995). Plots of k_La as a function of particle holdup are shown in Example Figure 8.3 for two impeller tip speeds and two superficial gas velocities.

In general, k_La decreased slightly as particle holdup increased, although not significantly. This small decrease contrasts with the slight increases in gas holdup observed with solid spherical particles (Example Fig. 8.2). The polyurethane foam BSPs appear to have a coalescing effect on gas bubbles resulting in the formation of larger bubbles. This causes the gas-liquid specific interfacial area to decrease, leading to the observed decrease in k_La. The effect is opposite of that observed with the spherical polyethylene particles which were observed to have a dispersing effect on gas bubbles. However, neither effect is particularly marked.

At larger particle holdups (> 20%, not shown in Example Fig. 8.3) the movement of the particles in the tank became very restricted. As a result this made it difficult for the air to rise to the liquid surface, leading to an

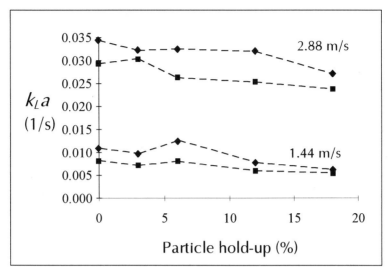

Example Fig. 8.3. Oxygen mass transfer coefficient as a function of particle holdup for two superficial gas velocities (■) 0.4 cm/s and (◆) 0.7 cm/s, and two impeller tip speeds (1.44 m/s and 2.88 m/s) in a 8.5 L STR.

increase in the residence time of the gas phase and to an increase in k_La value. However, such an observation is likely to be very scale-dependent and should not be assumed to be generally applicable.

CONCLUSION

It can be seen from the above results that oxygen transfer in the CBR is dependent on the circulating nature of the system, the presence of the neutral buoyancy particles and the porous nature of the particles. In the STR the influence of particles is considerably less, presumably because of the greater amount of energy being input by the mechanical agitator. It is likely, therefore, that if considerable increases in mass transfer can be brought about by decreases in bulk viscosity, through the use of immobilized cells, then the physical presence of the particles will not cancel out this effect.

ACKNOWLEDGMENTS

The material for this example was adapted from the MSc Thesis of Dr. Jonathan F. Dean (UMIST, 1985), the MSc Dissertation of Mr. Csaba Szabo (UMIST, 1988) and the PhD Thesis of Dr. Paulo A.L. Rodrigues (UMIST, 1995).

REFERENCES

1. Atkinson B. Immobilized cells, their applications and potential. In:Webb C, Black GM and Atkinson B, eds. Process Engineering Aspects of Immobilized Cell Systems. Rugby: IChemE Publ, 1986:3-19.

2. Thomas C. Image analysis: putting filamentous microorganisms in the picture. Trends Biotechnol 1992; 10:343-348.

3. Vasconcelos JMT and Alves SS: Direct dynamic k_La measurement in viscous fermentation broths—the residual gas holdup problem. Chem Eng J 1991; 47:B35-B44.

4. Pederson AG, Bundgaard-Nielsen M, Nielsen J, Villadsen J and Hassager O. Rheological characterization of media containing *Penicillium chrysogenum*. Biotechnol Bioeng 1993; 41:162-164.

5. Olsvik ES and Kristiansen B. On-line rheological measurements and control in fungal fermentations. Biotechnol Bioeng 1992; 40:375-387.

6. Buckland BC, Gbewonyo K, DiMasi D, Hunt G, Westerfield G and Nienow AW. Improved performance in viscous mycelial fermentations by agitator retrofitting. Biotechnol Bioeng 1988; 31:737-742.

7. Kawase Y and Moo-Young M. Mathematical models for design of bioreactors: Applications of Kolmogoroff's theory of isotropic turbulence. Chem Eng J 1990; 43:B19-B41.

8. Tucker KG, Mohan P and Thomas CR. The influence of mycelial morphology on the rheology of filamentous broths. Proc III Int Conf Bior Bioproc Fluid Dyn, BHR Group Conf Series Publ 5 1993; 261-273.

9. Gbewonyo K and Wang DIC. Enhancing gas-liquid mass transfer rates in non-Newtonian fermentations by confining mycelial growth to microbeads in a bubble column. Biotechnol Bioeng 1983; 25:2873-2880.

10. Wang DIC, Meier J and Yokoyama K. Penicillin fermentation in a 200-liter tower fermentor using cells confined to microbeads. Appl Biochem Biotechnol 1984; 9:105-116.

11. Robinson DK and Wang DIC. A transport controlled bioreactor for the simultaneous production and concentration of xanthan gum. Biotechnol Prog 1988; 4:231-241.

12. Van't Riet K. Mass transfer in fermentation. Trends Biotechnol 1983; 1:113-119.

13. Genon G, Manna L, Chiampo F, Sicardi S and Conti R. The problem of the oxygen transfer in industrial fermenters: Effect of the uneven gas distribution on the reactor performance. 6th Eur Congr Biotechnol Abstract Books 1993; I:MO110.

14. Olsvik ES and Kristiansen B. A k_La correlation for filamentous fermentations taking into account biomass concentration, growth rate, oxygen tension, rheological properties and the mixing qualities of the broth. 6th Eur Congr Biotechnol Abstract Books 1993; I:TU150.
15. Kristiansen B. Strategies in bioreactor modeling: Integration of kinetics and fluid dynamics. Proc Int Symp Bioreactor Perform, The Biotechnology Research Foundation 1993; 209-219.
16. Zhao XM, Gao Y, Hu ZD and Wang DZ. Citric acid fermentation in bioreactors with different impeller combinations. Proc III Int Conf Bior Bioproc Fluid Dyn, BHR Group Conf Series Publ 5 1993; 433-443.
17. Vardar F and Lilly MD. Effect of cycling dissolved oxygen concentrations on product formation in penicillin fermentations. Eur J Appl Microbiol Biotechnol 1982; 14:203.
18. Enfors SO and Larsson G. Microbial dynamics: The short term perspective.Proc Int Symp Bioreactor Perform, The Biotechnology Research Foundation 1993; 109-119.
19. Yasukawa M, Onodera M, Yamagiwa K and Ohkawa A. Gas hold-up, power consumption and oxygen absorption coefficient in a stirred tank fermentor under foam control. Biotechnol Bioeng 1991; 38:629-636.
20. Smith JJ, Lilly MD and Fox RI. Effect of agitation on *Penicillium chrysogenum*. Biotechnol Bioeng 1990; 35:1011-1023.
21. Harvey LM and McNeil B. Dynamics in Bioreactors. Proc 6th Eur Congr. Biotechnol 1994; 9:875-878.
22. Chisti Y and Moo-Young M. Impeller associated shear effects in fungal fermentations and implications for bioreactor design. Proc 6th Eur Congr Biotechnol 1994; 9:961-964.
23. Merchuk JC. Why use air-lift bioreactors? Trends Biotecnol 1990b; 8:66-71.
24. Merchuk JC. Shear effects on air-lift bioreactors. Adv Biochem Eng 1990a; 44:65-95.
25. Nienow AW. Introduction to fluid dynamics: Stirred bioreactors. Proc Int Symp Bioreactor Perform, The Biotechnology Research Foundation 1993; 33-46.
26. Dunlop EH and Ye SJ. Micro-mixing in fermentors: Metabolic changes in *Saccharomyces cerevisiae* and their relationship to fluid turbulence. Biotechnol Bioeng 1990; 36:854-864.
27. Dunn IJ and Heinzle E. Types of understanding in scaling down and up, as illustrated with an oxygen sensitive culture. Proc Int Symp Bioreactor Perform, The Biotechnology Research Foundation 1993; 189-202.

CHAPTER 5

IMPROVED PRODUCT YIELDS

It has repeatedly been demonstrated in the literature that cell immobilization may play an important role in increasing the yields of various cell metabolites. The increased product yields may be attributed to many factors, such as extended use of the biocatalytic activities of cells in the immobilized state, advantageous metabolic changes, or channeling the flow of material inside the cell through a particular pathway.

The most common reason for the increased product yields in immobilized cell systems is through channeling of the flow of mass away from cell mass synthesis. In most of these systems excessive cell growth is undesirable and this leads to the use of media which reduce the cell growth rate and, thus, the biomass yield. Identifying a suitable strategy for operating the bioreactor so that the yield of a particular metabolite is maximized or minimized is a challenging proposition. Sayles and Ollis[1] presented an excellent theoretical analysis of such an operating strategy, i.e. the periodic operation of immobilized cell bioreactors, using a transient mathematical model for growth and product formation. They remind us that this strategy originates from waste treatment where the objective is not to maximize the yield of a particular product but to achieve more complete waste treatment:

"Forced oscillation of immobilized cell systems is not a new operation. An early application is the rotating disk contactor for aerobic waste treatment. Here, microbial slime is grown on the surface of rotating disks, which alternately contact the wastewater and the air (oxygen) by rotating in the vertical plane through wastewater and air...Using this contacting pattern, the oxygen and wastewater are supplied to the organisms in a periodic manner. The higher growth rates achieved provide more complete waste treatment."

Sayles and Ollis also demonstrated that proper nutrient cycling may increase the average product yield by at least a factor of three in non-growth-related fermentations, relative to steady operation, without any significant sacrifice in the average total productivity. On the other hand growth-related fermentations were shown to lose significant total productivity under most cycling conditions, while the average product yield was fairly unchanged at all cycling rates.

In another study, periodic addition of nutrients in a continuous reactor fed with a glucose medium which supported the production of butanol but did not permit growth of alginate-immobilized *Clostridium acetobutylicum* cells resulted in a yield coefficient of butanol on glucose of 0.2 and a biomass to butanol ratio of 0.02. In a corresponding system with immobilized growing cells the ratio of biomass to butanol was 0.52 - 0.76 and the formation of butyric and acetic acid increased, thereby reducing the yield coefficient for butanol to 0.11.[2]

The addition of metabolic inhibitors in immobilized *Saccharomyces bayanus* and *Schizosaccharomyces pombe* increased the conversion of glucose to ethanol and, thus, the specific ethanol productivity.[3] Similarly the addition of acetate–which is an inhibitory

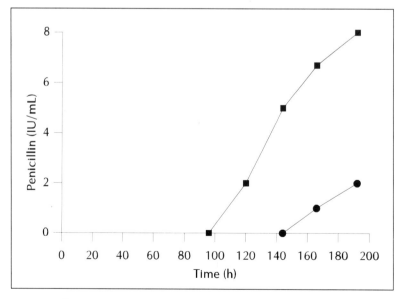

Fig. 5.1. Effect of cell immobilization on the penicillin yield by Penicillium chrysogenum in shaken flasks. (●) Free cell fermentation, (■) Immobilized cell fermentation.

by-product in ethanol fermentation–reduced cell overgrowth and increased ethanol productivity.

Amin et al[4] attributed the increased ethanol yields in their system not only to reduced cell growth but also to the formation of fewer by-products.

The carbon conversion efficiency of phosphate-limited immobilized *Streptomyces griseus* in candicidin formation was more than three-fold higher than that of growing cells.[5] The use of nitrogen-deficient medium and pure oxygen doubled the yield of citric acid in a dual hollow fiber reactor.[6]

In our laboratory, we have repeatedly observed improved product yields for a host of biological products and, especially, for non-growth associated products of the cell metabolism. Table 5.1 and Figures 5.1 and 5.2 are examples of increased antibiotic yields while the example box demonstrates the yield enhancement in the continuous production of cellulase by immobilized *Trichoderma viride*.

Figure 5.1 refers to the production of penicillin by *Penicillium chrysogenum* in shaken flasks, by both free and immobilized cells. The onset of penicillin production occurred at different times in the flasks. Concentrations of penicillin below 1.0 IU/mL were not detectable and thus the time of first detection of penicillin in a

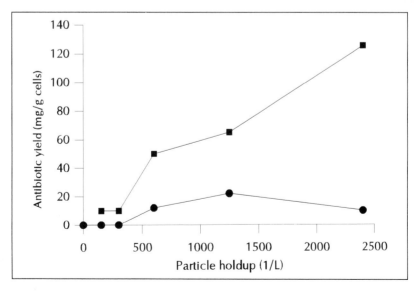

Fig. 5.2. Specific antibiotic yield of free and immobilized cells of S. longisporoflavus *in shaken flasks at different particle number densities. (●) Free cell fermentation, (■) Immobilized cell fermentation.*

Table 5.1. Yield of cells on lactose, penicillin yield on lactose and specific cell productivity after 192 hours

Flask	Cell Yield (g cells/g lactose)	Penicillin Yield (IU/g lactose)	Specific Productivity (IU/g cell.h)
Free Cells	0.41	0.18	2.35
Immobilized Cells	0.42	0.43	5.25

flask was in all probability not an indication of the onset of production. Production normally started after 40 hours of growth, however, where spore inocula were used a longer lag phase occurred and consequently the onset of production was delayed.

The highest level of production, 8.0 IU/mL, was achieved in flasks containing 60 ppi BSPs (Table 5.1). The specific productivity in the 60 ppi BSPs was 250% greater than in control shake flasks using freely suspended cells.

Figure 5.2 refers to the production of an intracellular antibiotic by *Streptomyces longisporoflavus* in shaken flasks for different particle holdups. The fact that the antibiotic was intracellular made the identification of the relative contributions of freely-suspended and immobilized cells fairly easy. It can be seen that the antibiotic yield of cells in the immobilized state was considerably higher than that of the freely-suspended ones, especially at high particle number densities. However, the total antibiotic yield of the immobilized cell system was only marginally better than that of the free system, due to the high proportion of penicillin-producing, freely-suspended cells in the former.

EXAMPLE 9:
ENHANCED CELLULASE YIELD
IN AN IMMOBILIZED CELL REACTOR

Secondary metabolites, such as cellulase, are usually produced under conditions of low or zero cell growth rate. This seriously limits the potential for continuous production, unless the cells can in some way be retained within the bioreactor at dilution rates in excess of their specific growth rate. Cell immobilization offers the possibility of maintaining this low or zero growth rate while operating continuously at high dilution rates, with no significant loss of cells from the bioreactor.

Materials and Methods

All experiments were performed using the filamentous fungus *Trichoderma viride* QM9123 and a glucose-based medium containing 10 mg/L a-D-sophorose monohydrate. The cells were immobilized in stainless steel BSPs manufactured by die pressing a length of knitted wire to form 6 mm diameter spheres having a porosity of 0.8.

Since the majority of cellulase-producing organisms are filamentous fungi, continuous production of the enzyme over extended periods by immobilized cells is normally hampered by the tendency of the mycelium to overgrow particles, leading to eventual blockage of the bioreactor. This problem has been alleviated by the development of a spouted bed configuration for the bioreactor (Example 10: Development of an Immobilized Cell Reactor for Use with Filamentous Fungi, chapter 6). In this type of reactor (Example Fig. 9.1), recycled liquid enters as a jet at the base of a bed of stainless steel BSPs, causing a column of the particles to spout upward. These are replaced by particles sliding down the sides of the vessel and into the path of the liquid jet. The resultant circulation of particles, through the region of high shear provided by the jet, ensures that any overgrowth of biomass is successfully removed by the abrasive forces of interparticle and particle-jet collisions and subsequently washed out of the bioreactor. In this way, continuous steady state operation of the bioreactor is achieved even under conditions where cell growth is occurring.

Following inoculation into an initial medium containing 35 g/L glucose, a period of batch operation was carried out in order to establish an immobilized cell population within the BSPs. Continuous operation was initiated after the residual glucose concentration had fallen to zero (usually at around 60 h after inoculation). During continuous operation fresh medium, containing between 2 and 3 g/L glucose, was added to the bioreactor at a dilution rate in excess of the maximum specific growth rate of the organism. Automatic addition of $2M$ KOH was used to control pH at 3.0.

Results and Discussion

A general trend observed in the experimental results is that cellulase production, commenced only after biomass accumulation had ceased and all glucose had been consumed. Cellulase concentration rapidly increased to a maximum value, which then remained stable for a considerable time. This latter observation may be of importance in the continuous production of cellulase on an industrial scale since the stability of the enzyme solution would enable its storage prior to processing. Data concerning the continuous production of cellulase by freely-suspended and immobilized cells, respectively, are presented in

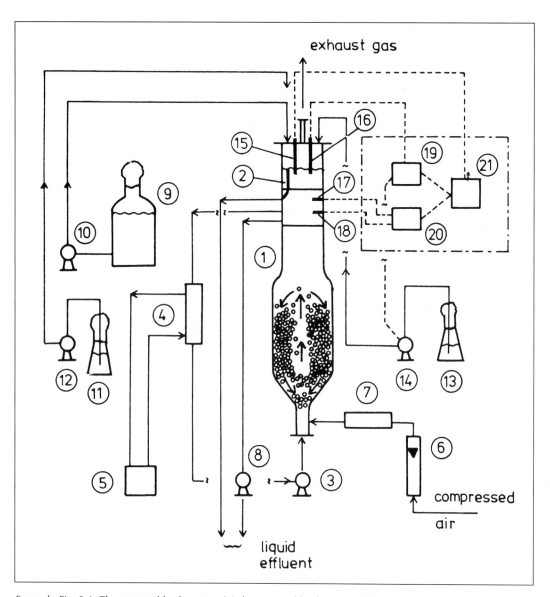

Example Fig. 9.1. The spouted bed system. *(1) the spouted bed reactor; (2) overflow pipe; (3) broth recycle pump; (4) heat exchanger; (5) cooler; (6) flow meter; (7) air filter; (8) liquid effluent pump; (9) medium reservoir; (10) medium feed pump; (11) antifoam storage bottle; (12) antifoam storage pump; (13) KOH reservoir; (14) KOH feed pump; (15) dissolved oxygen electrode; (16) pH electrode; (17) electric heater; (18) thermometer; (19) pH controller; (20) temperature controller; (21) chart recorder.*

Example Table 9.1. *Comparison of continuous cellulase production processes*

Process	Dilution rate (1/h)	Cellulase concentration (FPA U/L)	Bioreactor productivity (FPA U/L.h)	Yield (FPA U/g glucose)
Freely suspended cells[a]	0.012	900	10.8	36
Immobilized cells	0.15	210	31.5	96.8

[a] results taken from MA Zainudeen, PhD thesis, UMIST, 1974. FPA U (filter paper activity units) are the units of cellulase activity.

Example Table 9.1. It is clear from Example Table 9.1 that both the bioreactor productivity and the enzyme yield of the immobilized cells on glucose are substantially higher (almost three-fold) than the corresponding values for the freely-suspended cells.

The difference in productivity of the two continuous culture systems is clearly illustrated in Example Figure 9.2, which also shows data for specific productivity. The better performance of the immobilized cell system cannot be attributed to a superior organism/medium. The improvement in performance observed with the spouted bed/BSP system is clearly attributable to the presence of the immobilized cells.

An advantage normally claimed for immobilized cell systems is that of increased cell concentration within the reactor, leading to higher overall productivity, despite the often lower specific productivities which can result from diffusional limitations. It is interesting to note here that the high volumetric productivity shown in Example Figure 9.2 for the BSP system was obtained with a slightly lower overall cell concentration than that reported for the freely suspended cell system. The specific productivity of the immobilized cells was, therefore, considerably higher than that of the freely suspended cells, and it is apparent that, in this case, the presence of a diffusional limitation within the BSPs had a beneficial effect on cellulase production. A possible explanation for this is the low substrate concentration within a BSP, which results from the existence of a concentration gradient, and gives rise to an extremely low growth rate, which in turn induces cellulase production. This will be further explored in chapter 9 where a diffusion-reaction model is applied to estimate the concentration gradients within the BSPs used in the cellulase experiments. The low cell growth rate also leads to less of the substrate being used for biomass production and a consequent increase

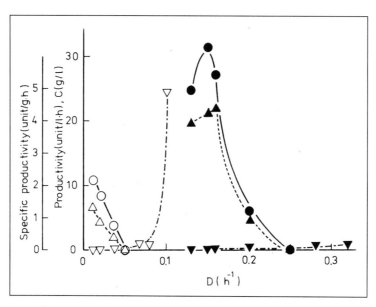

Example Fig. 9.2. Effects of dilution rate on cellulase productivity for freely suspended and immobilized cell systems. (O), bioreactor productivity for freely suspended cell system; (△), specific productivity for freely suspended cell system; (▽) glucose concentration for freely suspended cell system; (●), bioreactor productivity for immobilized cells system; (▲), specific productivity for immobilized cells system; (▼), glucose concentration for immobilized cells system. Data for freely suspended cells are taken from MA Zainudeen (Ph D Thesis, UMIST, 1974).

in cellulase yield. This represents a particular advantage of cell immobilization for secondary metabolite production since, in freely suspended cell systems, a considerable proportion of the substrate is required for biomass production, either during the growth phase of a batch culture or continuously during continuous culture.

CONCLUSION

Intra-particle substrate concentration gradients may lead to enhanced metabolite yields. Low cell growth rates within the particle may create favorable conditions for the production of non-growth associated metabolites and, therefore, improve their yields.

REFERENCES

1. Sayles GD and Ollis DF. Periodic operation of immobilized cell systems: analysis. Biotechnol Bioeng 1989; 34:160-170.
2. Foerberg C, Enfors SO and Haeggstroem L. Control of immobilized nongrowing cells for continuous production of metabolites. Eur J Appl Microbiol Biotechnol 1983; 17:143-147.
3. Amin G, Standaert P and Verachtert H. Effects of metabolic inhibitors on the alcoholic fermentation by several yeasts in batch or in immobilized cell systems. Appl Microbiol Biotechnol 1984; 19:91-99.
4. Amin G, Van den Eynde E and Verachtert H. Determination of by-products formed during the ethanolic fermentation using batch and immobilized cell systems of Zymomonas mobilis and Saccharomyces bayanus. Eur J Appl Microbiol Biotechnol 1983; 18:1-5.
5. Karkare SB, Venkatasubramanian K and Vieth WR. Design and operating strategies for immobilized living cell reactor systems. Part I. Biosynthesis of candicidin. Ann N Y Acad Sci 1986a; 469:83-90.
6. Chung BH and Chang HN. Aerobic fungal cell immobilization in a dual hollow-fiber bioreactor: continuous production of citric acid. Biotechnol Bioeng 1988; 32:205-212.

================ CHAPTER 6 ================

IMPROVED REACTOR CHOICE

Taking advantage of the beneficial consequences of immobilization can be as simple as adding an inert particulate support to a conventional bioreactor prior to sterilization. However, the majority of existing free-cell bioreactors are batch operated stirred tanks which are not particularly suited to continuous operation with particulate suspended solids. So, for maximum benefit, alternative bioreactor configurations are often suggested.

The removal of the dependency between reaction kinetics and residence time, together with the wide variety of forms that immobilized cells can take, should make an almost limitless range of bioreactor configurations possible. As new bioprocesses develop this versatility becomes of increasing importance as stressed more than a decade ago by Margaritis and Wallace:[1]

"New sophisticated bioreactor designs with unique performance characteristics will play a vital role in the economic manufacture of useful biotechnological products from natural and genetically modified cell systems of microbial, mammalian and plant origin."

In theory, a complete variation of the residence time distribution between plug-flow and well-mixed is possible in immobilized cell systems enabling better exploitation of the intrinsic reaction kinetics. For example, substrate limitation/product inhibition in continuous ethanol fermentations dictate the use of plug-flow reactors. Such operation is impossible with freely suspended cells but can be approached with immobilized cells through the use of fixed bed reactors. Bioreactors integrating production and separation are also easier to design due to the retention of the biomass in an immobilized cell system, irrespective of environmental perturbations (Example 11: Integrated Production and Separation Process (IPSEP), chapter 7).

REACTOR CONFIGURATIONS FOR IMMOBILIZED CELLS

A multiplicity of factors influence the choice of reactor type for a particular application. These include immobilization method, particle characteristics, e.g. shape, size, density, robustness, nature of substrate, inhibitory effects and hydrodynamic and economic considerations.[2] Immobilized cell reactors may be operated either batchwise or continuously, in plug flow mode or completely mixed. Completely mixed reactors can be advantageous if high substrate concentrations inside the bioreactor are to be avoided due to substrate inhibition though they are not particularly suitable for product inhibited reactions, as all cells are exposed to inhibitory levels of product.

The first rational development of an immobilized cell reactor was probably that of Pasteur, who proposed an acetifier where the *Acetobacter* cells were immobilized on the surface of wood chips. This type of biological reactor, the trickling filter, is still used today for vinegar production and for wastewater treatment and is shown schematically in Figure 6.1 along with other major categories of immobilized cell bioreactor.

STIRRED TANK REACTORS (STRs)

A major problem often referred to in arguments against the use of immobilized cell particles in STRs is the harsh treatment to which the particles are exposed. A high rate of shear may have severely damaging effects, especially in the case of gel particles. Modifications to classical STRs, e.g. enveloping the agitator in a porous mesh, can be made to allow mixing without destruction of the immobilized aggregates. The basket reactor shown in Figure 6.1, of which only laboratory applications are known, was developed in an attempt to protect catalyst particles from the disruptive action of the agitator. The catalyst is retained in mesh baskets that form the blades of the impeller. This gives good mixing and offers protection from attrition. However, as demonstrated in Example 4: Particle Stability in a Stirred Tank Reactor (chapter 2), such protection may not in fact be necessary.

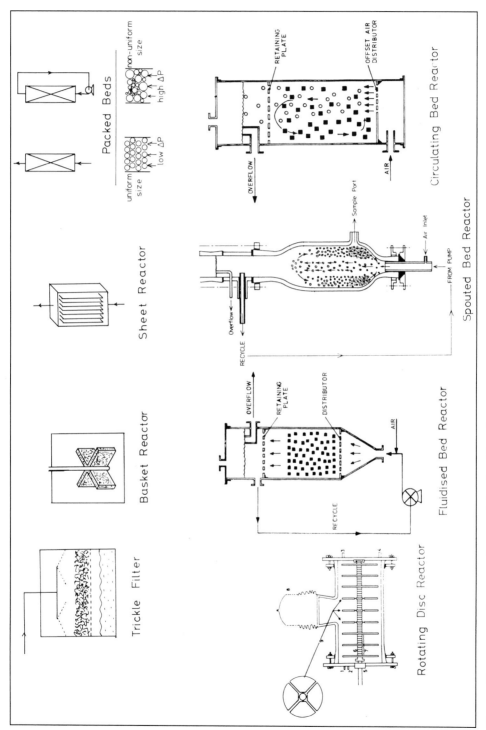

Fig. 6.1. Bioreactor configurations for immobilized viable cell systems.

FIXED BED REACTORS

Packed bed reactors are very common and, at least at the laboratory scale, convenient to use. They usually provide the tool for the first test of the applicability of cells immobilized by novel techniques. They are, however, difficult to scale-up and to quantify. Plug flow packed bed reactors operated on a once through basis may offer high rates of reaction due to high substrate concentration but have relatively poor mass and heat transfer coefficients due to low liquid velocities. Packed bed bioreactors with recycle show improved mass and heat transfer characteristics, plus improved controllability.

The size of the particulate carrier affects both pressure drop and internal diffusion resistance (Fig. 6.1). In well mixed packed bed reactors a compromise is sought for each process between pumping costs and mass transfer limitations. Qualitative factors like ease of handling may play a role in choosing the "best" particle size. The size distribution should be as uniform as possible since pressure drop depends upon bed voidage.

Particle compression caused by the static weight of the bed and the pressure due to flow is usually a severe drawback in packed beds of gel particles. Compression leads to a considerable decrease of the bed voidage and consequent increase in the pressure drop across the bed. The compaction of the bed can also drastically reduce the activity of the immobilized cells due to the consequent decrease in the specific surface area of the particles presented to the fluid.

Particulate substrates may preclude the use of a packed bed due to plugging and channeling problems. Packed bed reactors can also present considerable drawbacks in three phase operation. Gas-liquid contact is restricted, as is the release of gas from the voids of the bed. The release of gas is often observed to be slow, leading to gas accumulation in the form of stagnant slugs. This can cause gas flooding, which in turn produces poor liquid distribution and hence poor performance.[3]

Other forms of fixed bed reactor that alleviate these problems to some extent are the sheet reactor and the rotating disc reactor. In both systems, sheets of immobilized cells are typically arranged vertically and can operate in parallel flow mode or cross flow mode. The key difference between the two bioreactor types is the movement associated with the rotating disc reactor, which enables it to be operated with the discs half-submerged in the culture medium, moving alternately between liquid medium and air.

FLUIDIZED BED REACTORS (FBRS)

Fluidized bed reactors, in which particles become suspended as a result of the upflow of the fluid phase (usually a mixture of gas and liquid), have been the focus of much attention in the area of immobilized cell technology. Biological fluidized bed treatment of water and wastewater is a particular application of interest in which solid particles, e.g. sand, or porous particles (e.g. BSPs) can be used.[4] The use of FBRs has also been explored for aseptic systems.[5,6]

FBRs combine some of the advantages of STRs and packed beds and few of their drawbacks. Their attractive features include good mixing and mass transfer properties. In three phase operation gas-liquid contact and gas removal are facilitated as compared to packed beds. These are important characteristics for work with viable cells. Thus higher volumetric oxygen transfer coefficients can be attained and gas flooding avoided. Cell density per unit reactor volume is potentially less in a FBR than in a packed bed due to packing considerations. However, the overall performance may be higher in the FBR due to operational conditions.

An interesting aspect of the fluid mechanics of three phase fluidized systems is that gas holdup shows a marked variation with particle size, particle density and fluid flow rates. For instance, gas holdup and bubble size in beds of 1, 3 and 6 mm glass ballotini were measured and compared with those in the corresponding solids free system.[7,8] It was found that gas-liquid transfer is favored in beds of 6 mm particles due to bubble break-up and greater gas holdup. In this context there are reasons to justify consideration of particle size as a parameter in fluidized bed design, as well as concerning oneself with the biological limitations on particle size.

Particle density is also an important parameter, affecting pumping costs and mass transfer coefficients. The theoretical work of Riba[9] on mass transfer from a fixed sphere to a liquid in a fluidized bed suggested that the mass transfer coefficient depends upon the difference in density between particle and medium. This was confirmed by subsequent experimental studies.[10] It is therefore reasonable to suggest that the proper selection of the support matrix material may have beneficial effects on the rate of liquid-solid mass transfer in a fluidized bed. The density difference between a gel particle and the medium is usually very small. Immobilization in gels is nevertheless one of the most popular techniques for whole cell immobilization. Before attempting to use gel particles in a FBR it is important to be aware that whilst pumping costs will be low, the low density difference between the solid and liquid will not result in high rates of mass transfer. Furthermore, stable operation is difficult to obtain due to the small difference between the fluidizing velocity and the terminal settling velocity of the particles.

A specialized form of fluidized bed particularly suitable for use with immobilized mycelial fungi is the spouted bed bioreactor shown in Figure 6.1. The increased shear occurring at the base of the bed of dense particles provides for attrition of the tough mycelial tissue enabling steady control of biomass to be achieved (Example 10: Development of an Immobilized Cell Reactor for Use with Filamentous Fungi).

GAS AGITATED REACTORS

The use of gas to circulate the contents of a bioreactor through an external tube or internally, using a draft tube, is a convenient means (no rotating parts are involved) of achieving good mixing and aeration. Such "airlift" bioreactors are of simple construction and operation and are low power consumers. Consequently, they are very attractive for large scale operation and have been adopted by industry in a variety of processes using both free cells and cell aggregates, e.g. for cell mass production.[11]

The production at laboratory scale of cyclosporin A (an immunosuppressive agent) by the filamentous fungus *Tolypocladium inflatum* immobilized in carrageenan beads (4-5 mm) using an airlift reactor with external loop was reported by Foster et al.[12] The immobilization of the fungal organism was aimed at producing a less viscous liquid medium than that found in the submerged fermentation, facilitating the maintenance of adequate dissolved oxygen levels.

Black et al[6] devised the circulating bed reactor (CBR) shown in Figure 6.1, which facilitates the mixing of biomass carriers of essentially neutral buoyancy, e.g. polyurethane BSPs. Particle motion is induced by introducing air and/or recycled gas below the distributor. It was found that good liquid and particle mixing is most effectively achieved by introducing the gas over only a segment of the distributor. The CBR has been successfully operated with a number of immobilized viable cell applications (see examples throughout this book).

GENERAL CONSIDERATIONS

In general the requirements for an immobilized cell bioreactor are:

- a sufficiently low shear environment to preserve particle integrity
- maximal solids holdup
- sufficient local mixing to minimize dead zones and
- wherever possible, hydrodynamic conditions suited to the particular production kinetics

Substrate limitation usually means that this last requirement is best met by plug-flow conditions, particularly where product inhibition is also a feature of the production kinetics.

Conventional stirred tanks meet only the third of these requirements and are therefore of limited suitability. The majority of immobilized cell fermentation research is carried out in packed bed reactors. This is partly due to the origins of cell immobilization being in enzyme technology (where packed columns are ubiquitous) and partly due to the desire to operate plug flow systems. Although desirably simple and conceptually correct, packed beds suffer a number of limitations[13] and, in practice, rarely meet the third and fourth requirements. In situations where cell growth occurs, uncontrolled accumulation of biomass leading to channeling and blockage is a major problem. This is not so with fluidized beds, though these too fail to meet the third requirement unless relatively dense particles are employed. Most immobilized cell particles have net densities similar to that of the fermentation medium. For such particles, gas agitated systems such as the CBR are more appropriate, meeting all of the first three requirements. If the fourth requirement is for plug-flow then this necessitates, in theory, an infinite series of such bioreactors. However, in practice a close approximation might be achieved with just a few stages plus, perhaps, a final packed bed stage incorporated into a single bioreactor (Fig. 6.2). Such multistage systems may well be the bioreactors of the future.[14,15]

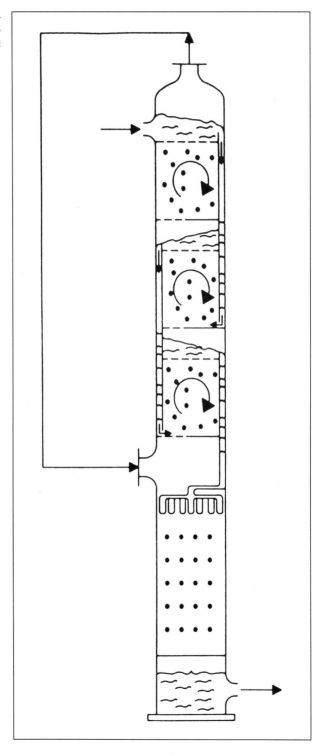

Fig. 6.2. Possible multi-stage bioreactor, featuring several CBRs and a packed bed.

EXAMPLE 10:
DEVELOPMENT OF AN IMMOBILIZED CELL REACTOR FOR USE WITH FILAMENTOUS FUNGI

In this example, some of the stages in the development of a laboratory scale immobilized cell bioreactor are described. Operating with filamentous fungi at any scale is difficult because of the ability of the hyphae to become attached and entangled with any available surface. At the laboratory scale this problem often makes extended, or continuous, operation impossible due to the inevitable blockages which occur as a result of such attached growth. During studies on citric acid production using *Aspergillus niger* immobilized in stainless steel BSPs, operational problems were encountered due to the filamentous nature of the organism. These were exacerbated by the fact that the cells were not only viable but growing very vigorously and it became necessary to rethink the design of bioreactor being used for the studies. This example illustrates, to some extent, the wide variation possible in bioreactor design when the cells are in the immobilized form.

THE FLUIDIZED BED REACTOR

Citric acid studies were started in the fluidized bed reactor (Example Fig. 10.1) which was the preferred bioreactor type for use with stainless steel BSPs. The bed of particles was held between two perforated stainless steel retaining plates. Air was fed to the culture through a porous, vyon, distribution plate situated below the lower retaining plate and liquid recycle entered the bioreactor between the distributor and retaining plate.

Growth of fungus on the retaining plates caused eventual blockage during continuous operation (Example Table 10.1). Wire mesh plates were used to replace the perforated type resulting in extended run time but with the same eventual problem. It was therefore decided to remove the retaining plates and vyon air distributor. However, without the upper retaining plate there was a very high risk of particles leaving the bioreactor and entering the recycle pump. In an attempt to prevent this an expanded section was introduced above the bed of particles. This successfully prevented particles passing into the recycle by reducing the liquid velocity at the top of the bed to less than the minimum fluidizing velocity. Air was introduced directly into the liquid recycle which now entered as a jet to the base of the reactor. The new configuration unfortunately gave rise to a new problem in that the particles at the top of the bed, being almost stationary, quickly overgrew with fungus and the bed became blocked. The non-productive volume of the bioreactor

Example Fig. 10.1. The fluidized bed reactor.

Example Table 10.1 Stages in the development of the spouted bed reactor

Bioreactor	Dimensions	Volume	Internals	Max. run time	Reason for failure
Fluidized bed	4" column	7 L	retaining plates & distributor	3.5 days	blockage of retaining plates
Fluidized bed	4" column, 6" expanded section	12 L	none	1 day	particle overgrowth - bed blockage
Draft tube	4" column 2" draft tube	8 L	draft tube	16 days	particle overgrowth (recycling into pump)
Draft tube	6" column 2" draft tube	10.5 L	draft tube & baffle	38 days	growth on draft-tube - blockage
Spouted bed	6" column	9 L	none	90 days	NONE

(i.e. the parts not occupied by the immobilized cells) was also substantially increased, hence reducing overall productivities.

To solve this problem it would be necessary to provide increased shear yet still prevent particle recycle without retaining plates. A possible solution was to make use of a draft tube.

THE DRAFT TUBE REACTOR

A draft tube was introduced so that a region of high shear could be provided at the base of the reactor. Fluidized particles passed upwards through the narrow inner tube and returned to the base of the reactor via the annulus (Example Fig. 10.2). Although an improvement on the fluidized bed reactor (Example Table 10.1), there was a tendency for some particles to 'jet' up through the draft tube and be entrained into the liquid recycle. In addition, there was now an insufficient recirculation rate for particles as they passed around the outside of the draft tube, which again resulted in eventual blockage of the bed. A potential solution was to increase the ratio of external to internal tube diameters from 2:1 to 3:1 and to introduce a baffle above the draft tube to deflect particles down into the annulus.

By increasing the external diameter of the bioreactor while retaining the same draft tube, better circulation of the particles was achieved. The

Example Fig. 10.2. The draft tube reactor.

Example Fig. 10.3. The spouted bed reactor.

baffle above the draft tube also prevented entrainment of particles into the liquid recycle, even at very much increased flowrates. Notwithstanding the improved performance, the internal surfaces associated with the draft tube provided a base for fungal attachment and this effectively prevented indefinite operation. Clearly the solution to this problem would be to remove the draft tube!

THE SPOUTED BED REACTOR

The large diameter vessel was retained from the later draft tube reactor and a fine nozzle was introduced for the liquid recycle return. In this way it was possible to operate the bioreactor in a similar way to the draft tube bioreactor by judicious positioning of the liquid inlet. The recycled liquid enters as a jet at the base of the bed, causing a column of the particles to spout upward. These are replaced by particles sliding down the sides of the vessel and into the path of the liquid jet. The resultant circulation of particles, through the region of high shear provided by the jet, ensures that any overgrowth of biomass is successfully removed by the abrasive forces of interparticle and particle-jet collisions and subsequently washed out of the bioreactor. In this way, continuous steady state operation of the bioreactor could be achieved even under conditions where vigorous cell growth was occurring (Example Table 10.1).

This so-called spouted bed reactor (Example Fig. 10.3) developed for citric acid production using *Aspergillus niger* was also successfully used for the production of cellulase using immobilized cells of *Trichoderma viride* (Example 9: Enhanced Cellulase Yield in an Immobilized Cell Reactor, chapter 5).

REFERENCES

1. Margaritis A and Wallace JB. Novel bioreactor systems and their applications. Chem and Biochem Eng 1984; 2:447-453.
2. Fonseca MM da, Black GM and Webb C. Reactor configurations for immobilized cells. In: Webb C, Black GM and Atkinson B, eds. Process Engineering Aspects of Immobilized Cell Systems. Rugby: IChemE publ, 1986:63-74.
3. Ghose TK and Bandyopadhyay KK. Rapid ethanol fermentation in immobilized yeast cell reactor. Biotechnol Bioeng 1980; 22:1489.
4. Cooper PF and Atkinson B. Biological fluidized bed treatment of water and wastewater. Ellis Horwood, 1981.
5. Oda G, Samejima H and Yamada T. Continuous alcohol fermentation technologies using immobilized yeast cells. Proceedings of Biotech '83, 1983:587.
6. Black GM, Webb C, Matthews TM and Atkinson B. Practical reactor systems for yeast cell immobilization using biomass support particles. Biotechnol Bioeng 1984; 26:134.
7. Michelson ML and Ostergaard K. Hold-up and fluid mixing in gas-liquid fluidized beds. Chem Eng J 1970; 1:37.
8. Ostergaard K. Hold-up, mass transfer and mixing in three phase fluidisation. AIChE Symp Ser No 176 1978; 74:82.
9. Riba JP. These de Doctorat d'Etat. Institut National Polytechnique de Toulouse 1978.
10. Riba JP, Routie R and Couderc JP. Fluidisation. Cambridge University Press 1978; 157.
11. Scott R. The design and evaluation of experiments to prove scale-up data. Proceedings of Biotech '83 1983; 235.
12. Foster BC, Coutts RT, Pasutto FM and Dossetor JB. Production of Cyclosporin A by Carrageenan—Immobilized Tolypocladium Inflatum in an Airlift Reactor with External Loop. Biotechnol letters 1983; 5:693.
13. Nunez MJ and Lema. Cell immobilization: Application to alcohol production. Enzyme Microb Technol 1988; 9:642-651.
14. Webb C, Dervakos GA and Dean JF. The importance of mixing in immobilized cell systems. Proc 2nd Int Conf Bioreactor Fluid Dynamics 1988; 199-213.
15. Klein J and Kressdorf B. Rapid ethanol fermentation with immobilized *Zymomonas mobilis* in a three stage reactor system. Biotechnol lett 1986; 8:739-744.

======= CHAPTER 7 =======

OTHER PROCESS IMPROVEMENTS

"Given the greatly increased flexibility that the biomass support particles bring to the whole approach to biological process engineering, it is not possible, even speculatively, to attempt to identify all the implications."

Atkinson et al[1]

IMPROVED DOWNSTREAM PROCESSING

The ability to treat cells as a discrete phase offers significant opportunities for facilitating the cell/liquid separation step. Recovering cells from the bioreactor either for disposal or for further processing can be achieved simply by draining the bioreactor, i.e. the first downstream processing step is integrated with fermentation. A further consequence of confining the cells as a discrete phase, dispersed throughout the fermentation medium, is the potential for maintaining lower bulk liquid viscosities than would prevail in normal free-cell culture. As described in chapter 4, this can be an important factor in sustaining reasonable gas-liquid mass transfer rates, particularly in fermentations involving filamentous organisms. It can also make cell/liquid separation much easier. Figure 7.1 shows that filtration of a *Penicillium chrysogenum* broth is considerably easier when immobilized cells are used (P.A.L. Rodrigues, PhD Thesis, UMIST, 1995). The potential also exists for cell production and recovery to be carried out in the same vessel, thus integrating the first downstream processing step with fermentation. An example of such a system where fermentation and solid/liquid separation are carried out in the same vessel

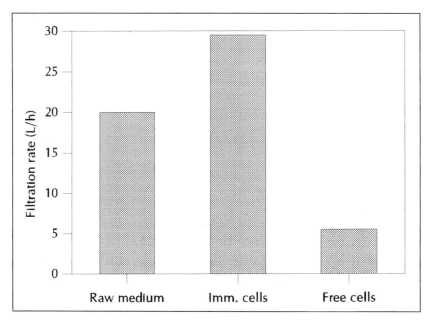

Fig. 7.1. Results of filtration tests on fermentation broths for free and immobilized cell penicillin fermentations, after 72 hours of fermentation.

is illustrated in Example 11: Integrated Production and Separation Process.

The recovery and purification of products may also be affected by cell immobilization. For example, κ-carrageenan immobilized growing cells of *Myxococcus xanthus* were reported to secrete foreign proteins out of the beads while native proteins were found to be retained predominantly within the gel; the beads represented a first purification step for protein recovery.[2] Cell immobilization has also been used in the separation of intracellular metabolites.[3] This separation was simplified by culturing the cells in membrane-covered alginate capsules, lysing the cells, and allowing the metabolites to diffuse into the extra-capsular medium, separated by molecular size.

The use of microencapsulation has been reported to facilitate the purification of monoclonal antibodies[4]: during culture the monoclonal antibodies secreted by the encapsulated hybridoma cells are retained within the capsule which allows the exit of low-molecular weight contaminating proteins. The capsules are then opened physically to harvest the antibody which can then account for up to 80% of the total protein present—a much higher concentration of antibody than is possible using other technologies.

Finally, co-immobilization of cells with magnetic particles may facilitate downstream processing in situations where particulate solids are encountered.

ADVANTAGEOUS PARTITION EFFECTS

The chemical composition of the carrier may play an important role in establishing a favorable microenvironment for the immobilized cells due to partitioning effects. The use of an appropriate matrix may increase the substrate concentration near the immobilized cells. Fukui and Tanaka[5] demonstrated the effectiveness of hydrophobic gels in bioconversions involving poorly water-soluble organic substrates, e.g. steroids. Ito et al[6] used Ca-alginate gel containing active carbon in order to raise the concentration of organic matter near the microbial cells and increase the efficiency of sewage treatment.

Similarly, the use of an appropriate matrix may decrease the product concentration near the immobilized cells. Nakajima et al[7] demonstrated the effectiveness of using a hydrophilic gel in pigment production: the partition coefficient of the poorly water-soluble pigments between gel/medium decreased with increasing gel hydrophilicity and this resulted in enhanced release of the pigments from the gels. Yeast cells immobilized in κ-carrageenan produced a very high concentration of ethanol (114 g/L) at a very high value of glucose conversion (> 0.95). It was suggested by the authors[8] that the ethanol shunned the very polar gel phase and thus reduced the extent of product inhibition within the aggregate.

IMPROVED PRODUCT STABILITY

The often high liquid throughputs in immobilized cell reactors often result in short residence times for various unstable products and, therefore, less pronounced degradation. An interesting situation arises when the product is degraded by enzymes which are produced by the cells during the fermentation. It may be possible that a suitably selected carrier will selectively entrap the degrading enzymes and allow the desired product to be excreted into the medium. This has been demonstrated in the production of α-mating factor with growing yeast cells immobilized in a photocrosslinkable resin.[9] The opposite is also possible; proteases outside, product inside.[4]

ADVANTAGES DUE TO CELL PROXIMITY

The close proximity of cells within an immobilized cell aggregate is important for some eukaryotic cells which have the capability to communicate. It is also advantageous when different species are co-immobilized.

Reactions which are normally performed in a series of reactors can now be carried out in the same vessel with significant productivity/cost advantages. An example of such a system is the conversion of starch to ethanol which involves an enzymatic step followed by a fermentation step: using co-immobilization, both the saccharification of starch and the subsequent ethanol fermentation can be carried out in the same reactor using various co-immobilized biocatalysts, e.g. glucoamylase/*Saccharomyces cerevisiae*,[10] glucoamylase/*Zymomonas mobilis*[11] and *Aspergillus awamori/Z. mobilis*.[12]

Algae have been used in co-immobilized systems to provide oxygen or reduced NADP for heterotrophic components. The co-existence of cells within a matrix may also enhance the yield of certain metabolic products. This has been demonstrated by Egorov et al[13] in a system containing co-immobilized *Nocardia* and *Arthrobacter* species, which produced much higher fibrinolytic proteinase activity than immobilized *Nocardia*. It was found that the *Arthrobacter* species—which does not produce fibrinolytic proteinase—synthesizes a polysaccharide substance which stimulates proteinase synthesis by *Nocardia*.

The close proximity of immobilized cells provides an alternative technique for conjugation. The results of Steenson et al[14] demonstrated efficient conjugal transfer of plasmid DNA among alginate-immobilized streptococcal cells and suggested that this method could be used as an alternative to conventional solid-surface and filter matings with these organisms.

REACTION SELECTIVITY

The different permeabilities of various substances in a complex reaction medium may be exploited in such a way that the selectivity of a particular reaction is increased. For example, intact cells of *Rhodotorula rubra* indiscriminately consumed all the constituents of an enzymic hydrolysate of a whey protein preparation, but cells entrapped in a polypyrrole membrane matrix consumed only low-molecular-weight components to give a low-phenylalanine peptide mixture in a high yield.[15]

EXAMPLE 11:
INTEGRATED PRODUCTION
AND SEPARATION PROCESS (IPSEP)

In the IPSEP system, fermentation and separation are carried out sequentially in the same vessel, a feature which can be particularly useful where the maintainence of sterile conditions is desired. In addition, IPSEP enables the recovery of biomass from the immobilization particles which can be recycled to the bioreactor for further use. During the fermentation, microbial cells grow into the reticulated polyurethane foam BSPs. Separation is carried out at the end of fermentation. First the 'bulk' liquid of nearly cell-free content is removed from the bioreactor thus separating the immobilized cells from the broth. Then the immobilized cells are recovered from the BSPs as product or for further processing. The cleaned BSPs after cell recovery can thus be reused. The integrated fashion of the IPSEP system not only greatly reduces the process complexity but also minimizes the risk of contamination. A prototype IPSEP unit is illustrated in Example Figure 11.1.

The IPSEP unit comprises several components: base plate, fermentation column, top plate, a threaded central shaft attached to a piston and recovery plate and a guide rod. In a typical fermentation, cells are immobilized into BSPs. For aerobic fermentations, air can be introduced through holes in one side of the piston; in this way the BSPs can be circulated and the liquid can be aerated. Medium is drained at the end of the fermentation and the recovery process can then commence. The central shaft is rotated, raising the piston to compress the particles against the recovery plate. In this way the immobilized cells are separated from the BSPs. The guide rod is positioned to prevent the piston from spinning when the shaft is rotated. The piston can move either upwards or downwards, depending on the direction of rotation of the central shaft.

Key to the effectiveness of IPSEP is the efficient separation of cells from the support particles. This depends not only on the mechanical properties of the BSPs (compressibility, reversibility of deformation) but also on the technique used to de-immobilize the cells (e.g. compression, liquid turbulence, vibration).

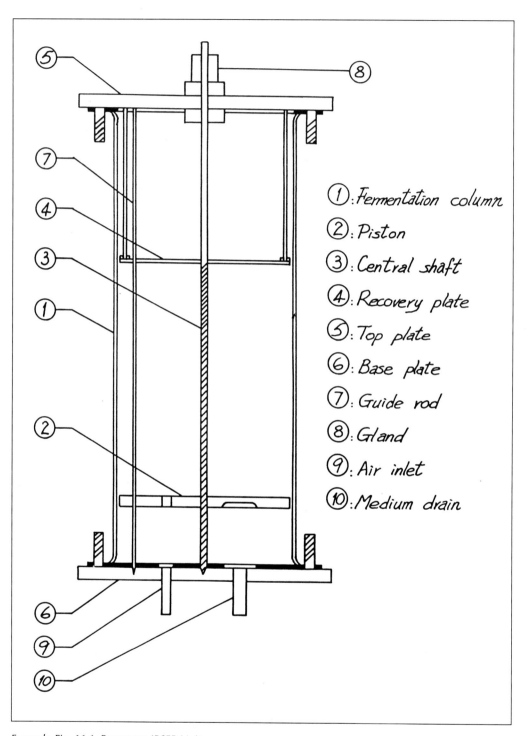

1 : Fermentation column
2 : Piston
3 : Central shaft
4 : Recovery plate
5 : Top plate
6 : Base plate
7 : Guide rod
8 : Gland
9 : Air inlet
10 : Medium drain

Example Fig. 11.1. Prototype IPSEP Unit.

The Effect of Direct Compression on Cell De-immobilization in the IPSEP Unit

The first cell/BSP separation-by-compression experiment yielded rather poor results, the cell recovery being only 27%. However, after introducing hydrochloric acid into the BSP bed, deflocculation of cells took place immediately and a milky solution was observed. The BSPs appeared very clean after this milky solution was removed from the column. Cell recovery now reached 85%.

The Effect of Pulsed Compression on Cell De-immobilization

To improve the separation efficiency, beds of BSPs containing immobilized cells were subjected to repeated pulses of compression. During the pulsed motion of the plunger, either previously drained medium or distilled water was drawn in and out of the bed separately. It was observed that the liquid being ejected from the syringe became more and more cloudy with yeast. This suspension contained no cell flocs indicating that the yeast had already been deflocculated prior to being separated from the BSPs. This was attributed to the combination of liquid turbulence and oscillating pressure caused by the pulsed motion. When water was sucked into the syringe, the turbulence was sufficient to break down the intercellular bonding, thus deflocculating, the cells. This allowed them to come out from BSPs when additional compression was provided. Cell recovery, however, was not particularly high (32%).

The Effect of Liquid Turbulence on Cell De-immobilization

To assess the effect of liquid turbulence alone on cell/BSP separation, distilled water was sucked into a packed bed of BSPs containing immobilized cells, using a vacuum pump. Prior to this the excess medium within the bed was allowed to drain out from the column and then the bed was gently squeezed to remove interstitial medium. As the vacuum pump was switched on and off, it was observed that the liquid retained in the column became turbid, more so than that observed previously. After finally switching off the pump, the liquid was allowed to drain from the column. Then the bed was squeezed to remove further liquid. Squeezing the bed after allowing the liquid to drain out did not appear to increase the cell concentration in the liquid. The efficiency of cell/BSP separation caused by liquid turbulence was about 55%. This efficiency, when compared with those obtained from direct compression (27%) and pulsed compression (32%), indicated that liquid turbulence alone was significant in improving cell/BSP separation.

THE EFFECT OF VIBRATION ON CELL DE-IMMOBILIZATION

The previous experiments had shown that some form of turbulence within the bed increased separation efficiency, apparently by disrupting cell flocs within the BSPs. Preliminary studies indicated that vibration of the BSP bed was also a promising method. After only a short period of time being vibrated on a rotary mixer, significant cell/BSP separation was observed. During the period of vibration, BSPs held in a wire basket inside a test tube were vibrated and spun simultaneously owing to the type of vibration supplied. After vibration, BSPs were rinsed with distilled water to wash out the cells. It appeared that the vibration force disrupted the yeast flocs in a similar way to liquid turbulence, whereas the constant rotation provided a centrifugal force, causing the cells to leave the BSPs, forming a thick layer of cells around the basket. Cell recovery was less if the BSPs were not gently squeezed prior to being vibrated. This was attributed to momentum being lost to the liquid, thus decreasing that reaching the cells. In this way the net effect of this type of vibration would be reduced. Hence in the following experiments, BSPs were gently squeezed to dislodge the interstitial liquid before commencing the vibration procedure. It was shown that either increasing the vibration speed or the vibration time increased the efficiency of cell recovery from BSPs. This was attributed to more frequent collisions between cells in flocs. The efficiency of cell recovery was also improved by increasing the tube diameter and decreasing the height of the BSP packed bed. This was caused by the position of the vibration source. Since the vibration source was at the bottom of the test tube, the lower portion of that was subjected to stronger centrifugal force. The BSPs in the lower portion of the test tube appeared to be freer of cells than those in the upper portion

Example Table 11.1. Effectiveness of mode of operation on cell deimmobilization in the IPSEP unit

Mode of operation	Cell recovery
Direct compression	27%
Direct compression + HCl	85%
Pulsed compression	32%
Liquid turbulence	55%
Vibration	90%

after vibration. It was also observed that by decreasing the amount of water used for washing, the separation efficiency was improved. Finally, by repeating the vibration and washing with distilled water cycle, the separation efficiency reached 100%. A typical average cell recovery in this mode of operation was 90%. The results from the above experiments are summarized in Example Table 11.1.

CONCLUSION

Introducing immobilized cells into the reactor means that, at some stage, we may have to de-immobilize them. De-immobilizing microbial cells has been largely ignored in the literature. The development of IPSEP demonstrates that the resilience of BSPs makes them ideal for such a task, especially under conditions of liquid turbulence and vibration.

REFERENCES

1. Atkinson B, Black GM and Pinches A. Process intensification using cell support systems. Process Biochemistry 1980; 15:24-32.
2. Younes G, Breton AM and Guespin-Michel J. Production of extracellular native and foreign proteins by immobilized growing cells of *Myxococcus xanthus*. Appl Microbiol Biotechnol 1987; 25:507-512.
3. Jarvis AP Jr. Recovery and purification of a substance developed but not excreted by cells by encapsulation and lysis of the cell membrane". Fr Demande FR 2544330 A1 1984.
4. Duff RG. Microencapsulation technology: a novel method for monoclonal antibody production. Trends Biotechnol 1985; 3:167-170.
5. Fukui S and Tanaka A. Bioconversion of lipophilic or water insoluble compounds by immobilized biocatalysts in organic solvent systems. Enzyme Eng 1982b; 6:191-200.
6. Ito H, Tatsumi S and Nagai K. Fermentation with immobilized microbial cells". JP 61/104792 1986.
7. Nakajima H, Sonomoto K, Morikawa H, Sato F, Ichimura K, Yamada Y and Tanaka A. Entrapment of *Lavandula vera* cells with synthetic resin prepolymers and its application to pigment production. Appl Microbiol Biotechnol 1986; 24:266-270.

8. Wada M, Kato J and Chibata I. Continuous production of ethanol in high concentration using immobilized growing yeast cells. Eur J Appl Microbiol Biotechnol 1981; 11:67-71.

9. Okada T, Sonomoto K and Tanaka A. Application of entrapped growing yeast cells to peptide secretion system. Appl Microbiol Biotechnol 1987; 26:112-116.

10. Fukushima S and Yamade K. Continuous alcohol fermentation of starch materials with a novel immobilized cell/enzyme bioreactor. Ann NY Acad Sci 1984; 434:148-151.

11. Rhee SK, Lee GM, Kim CH, Abidin Z and Han MH. Simultaneous sago starch hydrolysis and ethanol production by *Zymomonas mobilis* and glucoamylase. Biotechnol Bioeng. Symp 1986; 17:481-493.

12. Tanaka H, Kurosawa H and Yajima M. Fermentation by immobilized anaerobic and aerobic microorganism cells". Jpn Kokai Tokkyo Koho JP 61/209590 A2 1986.

13. Egorov NS, Landau NS and Kotova IB. Immobilized systems producing fibrinolytic proteinases. Dokl Akad Nauk SSSR 1987; 296:475-477.

14. Steenson LR and Klaenhammer TR. Conjugal transfer of plasmid DNA between streptococci immobilized in calcium alginate gel beads. Appl Environ Microbiol 1987; 53:898-900.

15. Watanabe M, Matsumura M, Yabuki S, Aizawa M and Arai S. Construction of a bioreactor with immobilized yeast cells for production of a low-phenylalanine peptide mixture as a phenylketonuria foodstuff. Agric Biol Chem 1988; 52:2989-2994.

=========CHAPTER 8 =========

POTENTIAL DISADVANTAGES

The key difference between immobilized and non-immobilized cells systems is, of course, the heterogeneity brought about by the introduction of particles into the bioreactor. Such heterogene potential disadvantages are the basis of many of the advantages associated with immobilized cells which have been explored in previous chapters. They are, however, also the main cause of potential disadvantages. On the subject of heterogeneity and bioprocess scale-up, industrialists[1] have recently concluded that:

"...research on the scale-up of heterogeneous bioprocesses is mainly hampered by lack of available measuring methods and experimental data. As tools have to be developed, as the phenomena are complex and as only limited experimental data are available at present it is to be expected that in this research area no major breakthrough will take place in the near future."

The problems introduced by immobilization, be they operational, biological or mass transfer problems, can, in principle, be solved. Judicious selection of reactor operating conditions and the use of particles with improved physico-chemical properties often suffice to alleviate most of these problems. The real problem, which may be much more difficult to address, is the lack of expertise of the biotechnological industry in large scale immobilized cell processes. Understandably, industrialists do not like poorly-understood and poorly-controlled heterogeneous systems. However, the first industrial applications have already emerged[2] and this is bound to create a virtuous circle for cell immobilization technology.

MECHANICAL PROBLEMS

Although the catalytic stability of immobilized cells is often improved compared to freely-suspended cells, the rate limiting step in the operational stability of these systems may be the stability of the carrier. A number of different mechanisms, discussed below, are responsible for the reduced stability of the immobilization carrier.

INTRA-PARTICLE PHASE FORMATION

When the concentration of a gaseous product within an immobilization matrix exceeds its saturation value then intraparticle gas formation will occur which may lead to, for example, the flotation and/or rupture of the beads. This has been observed repeatedly in ethanol fermentations where carbon dioxide is one of the main products. Intraparticle gas formation can be eliminated by enhancing the mass transfer rate of the gas out of the particle. This can be achieved by using small, non-densely populated particles and by minimizing the gas holdup in the bulk of the aqueous phase. The latter can be accomplished by using special reactor configurations such as inclined or horizontal or compartmentalized reactors.[3] It should be noted that the high carbon-dioxide holdup in some immobilized cell systems may lead to back-mixing, which is undesirable for plug-flow operations, and even to heat transfer problems.[4] Solid or liquid phase formation is also possible within an immobilized cell particle, i.e in the production of poorly water soluble pigments or steroids.

CELL OVERGROWTH

Uncontrolled growth of immobilized cells apart from decreasing product yields, may also result in a number of operational problems. In adsorbed cell systems, bioreactor plugging is often observed as a result of cell overgrowth. The bioreactor plugging by cells of *Bacillus amyloliquefaciens* adsorbed on an ion exchange resin for α-amylase production was avoided by periodically substituting the nutrient medium with medium lacking in both starch and yeast extract.[5] The blockage of vertical or inclined packed beds in Acetone-Butanol-Ethanol fermentations by *Clostridium acetobutylicum* immobilized onto bonechar[6] was alleviated when a partially-packed bed reactor was operated in the horizontal mode, while a fluidized bed reactor was the most stable of the systems investigated. In dual hollow-fiber reactors, an unlimited cell growth may expand the fiber and compress the inner fibers to reduce the substrate flowrates gradually to zero. This has been observed for aerobic growth of *Escherichia coli*,[7] citric acid production by *Aspergillus niger*[8] and rifamycin B conversion by *Humicola* species.[9] The main parameter for control of cell growth was the nitrogen concentration in the medium. Addition of antibiotics which inhibit cell division but allow protein synthesis to take place has also been used to prevent the cells from clogging the matrixes in which they are embedded.[10]

THE MECHANICAL STABILITY OF THE MATRIX

The mechanical properties of the immobilization carrier are very important in determining the operational lifetime of an immobilized cell bioreactor. Mechanically-stable carriers can withstand the pressure exerted by the embedded growing cells due to intraparticle gas formation and are also able to resist harsh hydrodynamic conditions as well as compaction in packed beds. On the other hand, poor mechanical properties of an immobilization carrier may restrict substantially the choice of equipment as well as the choice of operating conditions.

The choice of the reactor and of the reactor operating conditions may be restricted by the mechanical stability of the carrier. Stirred tanks have been reported to damage gel beads. The use of an airlift reactor in the transformation of steroids promoted the stability of Ca-alginate beads in comparison to a stirred tank reactor.[11] Similarly a packed-column reactor gave considerably better ethanol productivity than a stirred tank reactor and this was partly attributed to the attrition of yeast cells and the disintegration of the gel particles.[12] A conical bubble column bioreactor preserved the activity of penicillin producing *Penicillium chrysogenum* for more than seven days while cells lost activity after two days in a simple bubble column. This result was attributed to better mixing as well as higher mechanical stability of the gel beads in the conical column.[13] A stirred catalytic basket reactor with immobilized yeast cells gave considerably better volumetric productivities than that of conventional stirred tank reactors.[14] Shirai et al[15] cultivated hybridoma cells in an expanded bed bioreactor under mild flow conditions in order to reduce the destruction of the gel particles while

Himmler et al[16] used an airlift reactor where a slowly turning marine-type impeller was placed in the draft tube, instead of using gas. Oxygen was supplied on occasional demand by the gas sparger.

The use of certain nutrient concentrations in the media may also be dictated by the mechanical stability of the carrier. For example, alginate beads are disrupted or dissolved in the presence of cation chelating agents such as phosphate. Selection of an appropriate pH and phosphate concentration has been reported to yield mechanically stable particles.[17] Numerous other factors have been reported to mechanically stabilize Ca-alginate beads such as selection of alginate and Ca^{++} concentration,[18] selection of the initial biomass concentration in the matrix and selection of the incubation period in $CaCl_2$ solution[19] and washing of the beads with a trivalent cation.[20] In addition, the internal gelation of alginate,[21,22] the use of barium and strontium instead of calcium[23] and the use of pectate instead of alginate[24] may yield stronger gel particles.

Polyethyleneimine modified alginate was reported to enable immobilized cells to grow by dissolving the surrounding gel matrix; the dissolved polymer adduct is displaced peripherally and gelled again by the influx of calcium ion from the surrounding fermentation broth, retaining both cells and carrier polymer in the gel beads. Thus, the immobilized cells gain space for growth by expanding the carrier matrix.[25]

CELL LEAKAGE

Although some cell leakage from the immobilization matrix may be desirable in order to maintain a steady state biomass concentration, especially with growing cells, too much is undesirable and may lead to the cessation of activity. In certain processes, e.g. food processing, cell leakage into the product stream may be totally undesirable. The rate of cell leakage depends on the rate of cell growth as well as other factors. Wang et al[26] investigated the effect of various parameters on the cell leakage kinetics using as a model system yeast cells immobilized in κ-carrageenan. Increasing the shaking speed or local fluid shear greatly increased the rate of cell leakage. The leakage was influenced by K^+ so that by increasing the K^+ in the bulk solution, cell leakage could be reduced to 10% of the initial cell concentration. Bajpai et al[27] reported that supplementation of a glucose medium with calcium chloride influenced the rate of cell leakage. Cell leakage from the beads decreased with increases in CaCl$_2$ concentration up to 2.0 g/L, but there was no appreciable change in cell leakage beyond this concentration. Bar et al[28] suggested that the amount of cells adsorbed during a fermentation process is different from that found from adsorption isotherm data and this may be partly due to the mechanical agitation causing detachment of the cells from the carrier. The electron micrographs of Karkare et al[29] illustrate how the yeast cells populate the beads and escape into the fermentation liquor.

TOXICITY OF THE CARRIER
Although in earlier systems immobilized cells suffered from the toxicity of the carrier[30] current immobilization carriers seem to have generally little or no toxicity effects on the cells.

INCREASED SUBSTRATE LIMITATIONS
Substrate limitations have been invoked in several cases in order to explain the low efficiency of immobilized cell systems, though in certain cases such an imputation may not be correct. When the microorganism is aerobic and/or the reaction requires molecular oxygen, then it is most likely that oxygen will be the rate limiting substrate because of its low solubility. Attempts have been made therefore to increase the oxygenation of immobilized cells. These include:
- Increased oxygen spatial pressure[31]
- Oxygenation in an external loop[32]
- In situ oxygen production with hydrogen peroxide or by co-immobilization with algae[33]
- Use of a hydrophobic matrix
- Use of other oxygen carriers, such as haemoglobin, perfluorochemicals or organic solvents[34]
- Use of membrane oxygenators

INCREASED PRODUCT INHIBITION
As the product concentration increases with distance within the particle the effects of product inhibition become locally more pronounced and this may lead to a decrease in the reaction rate. Toda et al[35] observed that most of the ethanol produced in a horizontal bioreactor was produced by free cells in suspended or settled states. The relatively low ethanol production by the immobilized

yeast cells was attributed to the higher product inhibition of the fermentation rate within the support. In a Japanese patent[36] a two-stage continuous bioreactor is described consisting of *Saccharomyces cerevisiae* immobilized in Ca-alginate beads; in the first reactor small beads are used (< 1.8 mm) while in the second large beads are used. Although one would expect increased product inhibition problems in the second bioreactor, in fact the large beads gave the immobilized microorganism a higher tolerance to the product and hence the ethanol concentration increases. Similarly, the ethanol tolerance of *S. cerevisiae* immobilized in a compound containing polyvinylchloride and polyvinylpyrrolidone was markedly increased.[37]

Furthermore over the last few years, a number of membrane reactors have been proposed which overcome substrate and product diffusional resistances by forcing the nutrient solution through the cell mass and/or separate the product from the reaction mixture so as to reduce product inhibition. Such an example is the membrane 'sandwich' reactor where the biocatalyst is sandwiched between an ultrafiltration and a reverse osmosis membrane.

Efythymiou and Shuler[38] reported that operating a membrane entrapped yeast reactor by pressure cycling eliminates diffusional resistances, increases the ethanol productivity by almost an order of magnitude and also eliminates the need for periodic nutrient addition.

REFERENCES

1. Feijen J, Hofmeester JJM and Groen D. Scale up of heterogeneous bioprocesses. Progress in Biotechnology 1994; 9:919-926.
2. Kyotani S, Nakashima T, Izumoto E and Fukuda H. Continuous interesterification of oils and fats using dried fungus immobilized in biomass support particles. J Ferment Bioengineering 1991; 71:286-288.
3. Qureshi N, Pai JS and Tamhane DV. Reactors for ethanol production using immobilized yeast cells. J Chem Technol Biotechnol 1987; 39:75-84.
4. Ghose TK and Bandyopadhyay KK. Studies on immobilized *Saccharomyces cerevisiae*. II. Effect of temperature distribution on continuous rapid ethanol formation in molasses fermentation. Biotechnol Bioeng 1982; 24:797-804.
5. Groom CA, Daugulis AJ and White BN. Continuous a-amylase production using *Bacillus amyloliquefaciens* adsorbed on an ion exchange resin. Appl Microbiol Biotechnol 1988; 28:8-13.
6. Qureshi N and Maddox IS. Reactor design for the ABE fermentation using cells of *Clostridium acetobutylicum* immobilized by ad-

sorption onto bonechar. Bioprocess Eng 1988; 3:69-72.

7. Chang HN, Chung BH and Kim IH. Dual hollow-fiber bioreactor for aerobic whole-cell immobilization. ACS Symp Ser 1986; 314:32-42.

8. Chung BH and Chang HN. Aerobic fungal cell immobilization in a dual hollow-fiber bioreactor: continuous production of citric acid. Biotechnol Bioeng 1988; 32:205-212.

9. Hwang YB, Chung BH, Chang HN and Han MH. Biological conversion of rifamycin B by live *Humicola* sp. cells immobilized in a dual hollow fiber bioreactor. Bioprocess Eng 1988; 3:159-163.

10. Buelow L, Birnbaum S and Mosbach K. Production of proinsulin by entrapped bacteria with control of cell division by inhibitors of DNA synthesis. Methods Enzymol 1988; 137:632-636.

11. Kloosterman J and Lilly MD. An airlift loop reactor for the transformation of steroids by immobilized cells. Biotechnol Lett 1985; 7:25-30.

12. Lee TH, Ahn JC and Ryu DDY. Performance of an immobilized yeast reactor system for ethanol production. Enzyme Microb Technol 1983; 5:41-45.

13. El-Sayed AHMM and Rehm HJ. Continuous penicillin production by *Penicillium chrysogenum* immobilized in calcium alginate beads. Appl Microbiol Biotechnol 1987; 26:215-218.

14. Gamarra JA, Cuevas CM and Lescano G. Production of ethanol by a stirred catalytic basket reactor with immobilized yeast cells. J Ferment Techno 1986; 64:25-28.

15. Shirai Y, Hashimoto K, Yamaji H and Tokashiki M. Continuous production of monoclonal antibody with immobilized hybridoma cells in an expanded bed fermentor. Appl Microbiol Biotechnol 1987; 26:495-499.

16. Himmler G, Palfi G, Rueker F, Katinger H and Scheirer W. A laboratory fermentor for agarose immobilized hybridomas to produce monoclonal antibodies. Dev Biol Stand 1985; 60:291-296.

17. Dainty AL, Goulding KH, Robinson PK, Simpkins I and Trevan MD. Stability of alginate-immobilized algal cells. Biotechnol Bioeng 1986; 28:210-216.

18. Cheetham PSJ, Blunt KW and Bucke C. Physical studies on cell immobilization using calcium alginate gels. Biotechnol Bioeng 1979; 21:2155-2168.

19. Ogbonna JC, Amano Y and Nakamura K. Elucidation of optimum conditions for immobilization of viable cells by using calcium alginate. J Ferment Bioeng 1989; 67:92-96.

20. Rochefort WE, Rehg T and Chau PC. Trivalent cation stabilization of alginate gel for cell immobilization. Biotechnol Lett 1986; 8:115-120.

21. Johansen A and Flink JM. Influence of alginate properties and gel reinforcement on fermentation characteristics of immobilized yeast cells. Enzyme Microb Technol 1986a; 8:737-748.

22. Johansen A and Flink JM. Immobilization of yeast cells by internal gelation of alginate. Enzyme Microb Technol 1986b; 8:145-148.
23. Tanaka H and Irie S. Preparation of stable alginate gel beads in electrolyte solutions using barium and strontium. Biotechnol Tech 1988; 2:115-120.
24. Berger R and Ruehlemann I. Stable ionotropic gel for cell immobilization using high molecular weight pectic acid. Acta Biotechnol 1988; 8:401-405.
25. Joung JJ, Akin C and Royer GP. Immobilization of growing cells by polyethyleneimine-modified alginate. Appl Biochem Biotechnol 1987; 14:259-275.
26. Wang HY, Lee SS, Takach Y and Cawthon L. Maximizing microbial cell loading in immobilized-cell systems. Biotechnol Bioeng Symp 1982; 12:139-146.
27. Bajpai PK, Wallace JB and Margaritis A. Effects of calcium chloride concentration on ethanol production and growth of immobilized *Zymomonas mobilis*. J Ferment Technol 1985; 63:199-203.
28. Mattiasson B and Hahn-Haegerdal B. Microenvironmental effects on metabolic behavior of immobilized cells. A hypothesis. Eur J Appl Microbiol Biotechnol 1982; 16:52-55.
29. Karkare SB, Dean RC Jr and Venkatasubramanian K. Continuous fermentation with fluidized slurries of immobilized microorganisms. Bio/Technology 1985; 3:247-251.
30. Starostina NG, Lusta KA and Fikhte BA. Morphological and physiological changes in bacterial cells treated with acrylamide. Eur J Appl Microbiol Biotechnol 1983; 18:264-270.
31. Ghommidh C, Navarro JM and Durand G. Acetic acid production by immobilized *Acetobacter* cells. Biotechnol. Lett 1981; 3:93-98.
32. Okuhara A. Vinegar production with *Acetobacter* grown on a fibrous support. J Ferment Technol 1985; 63:57-60.
33. Adlercreutz P and Mattiasson B. Oxygen supply to immobilized cells: 1. Oxygen production by immobilized *Chlorella pyrenoidosa*. Enzyme Microb Technol 1982; 4:332-336.
34. Adlercreutz P and Mattiasson B. Oxygen supply to immobilized cells: 3. Oxygen supply by hemoglobin or emulsions of perfluorochemicals. Eur J Appl Microbiol Biotechnol 1982; 16:165-170.
35. Toda K, Ohtake H and Asakura T. Ethanol production in horizontal bioreactor. Appl Microbiol Biotechnol 1986; 24:97-101.
36. Koga K, Koshimizu S, Okazaki M and Yasuba Y. Continuous fermentor". Jpn Kokai Tokkyo Koho JP 61/78374 A2 1986.
37. Ooshima H, Saitoh S and Harano Y. Preparation of immobilized yeast cells for production of highly concentrated ethanol. Kagaku Kogaku 1984; 48:368-370.
38. Efthymiou GS and Shuler ML. Elimination of diffusional limitations in a membrane entrapped cell reactor by pressure cycling. Biotechnol Prog 1987; 3:259-264.

======= CHAPTER 9 =======

DIFFUSION AND BIOLOGICAL REACTION: PRACTICAL APPLICATIONS OF THE CONVENTIONAL THEORY

It may appear rather odd to devote a whole chapter to discuss the practical applications of conventional diffusion-reaction theory in the context of immobilized viable cells, when one of the main theses of this book is that this theory is often inadequate to describe the behavior of such systems. Nonetheless, conventional theory, if applied with caution and if not taken at face value, is still invaluable, especially at preliminary design stages. Significantly better than order-of-magnitude arguments, but still nowhere nearer to providing an accurate representation of reality, conventional theory can provide guidance in such mundane yet extremely important matters as the selection of the most appropriate particle size and particle biomass holdup for a particular application. Much more importantly, conventional diffusion-reaction theory can explain the interplay between biological reaction and diffusive mass transport in immobilized cell aggregates. The latter is nicely summarized by Weisz[1] in his interdisciplinary excursion in the area of diffusion and chemical transformation:

"The chemical transformation of molecules takes place in an endless number of systems, natural and man-made. The rate and effectiveness of these transformations are influenced by the availability of

reactants. As transformation occurs, the original concentration of the reactants is depleted and, nearly always, is replenished by some process of diffusion. The net attainable speed of reaction thus depends on the relative competition of reactivity and of the capability for diffusive flow in the relevant parts of the system... Only space limits the number of examples that could be enumerated from geology, agronomy or physiology to the dyeing of garments or the pot on the kitchen stove."

In this chapter, various biological systems are analyzed in light of diffusion and reaction interactions. A handful of heuristic rules, derived from computer simulations, is also proposed.

CONVENTIONAL THEORETICAL ANALYSIS

The coupling of diffusion and reaction in an immobilized cell aggregate is often described by the second order differential equation:

$$D_{e,s} r^{-n} \frac{d}{dr}\left(r^n \frac{dC_s}{dr} \right) = R_s\left(C_s\right)$$

The boundary conditions are applied at two different locations in space:

$$r = 0; \quad \frac{dC_s}{dr} = 0$$

$$r = r_a; \quad D_{e,s} \frac{dC_s}{dr} = \varepsilon_a k_{l,s}\left(C_{s,b} - C_{s,i}\right)$$

where

$D_{e,s}$ is the effective diffusivity of the substrate

C_s is the substrate concentration

r is the distance from the particle center

n is a geometry dependent parameter (n = 0, 1, 2 for planar, cylindrical and spherical geometry, respectively)

R_s is the reaction rate per unit of aggregate volume

$k_{l,\,s}$ is the external mass transfer coefficient for the substrate

$C_{s,\,b}$ is the substrate concentration in the bulk of the aqueous phase

$C_{s,\,i}$ is the substrate concentration at the liquid-solid interface

ε_a is the porosity of the aggregate

r_a is a characteristic aggregate length (e.g. radius of a sphere)

The above diffusion-reaction equation is often reduced to the following dimensionless form:

$$\nabla^2 w_s = \phi_s^2 g_s \big[w_s(z) \big]$$

with boundary conditions

$$z = 0; \qquad \frac{dw_s}{dz} = 0$$

$$z = 1; \qquad \frac{dw_s}{dz} = Bi_s \big[1 - w_s(1) \big]$$

where

$$\nabla^2 w = z^{-n} \frac{d}{dz} \left(z^n \frac{dw_s}{dz} \right)$$

$$w_s = \frac{C_s}{C_{s,b}}$$

$$z = \frac{r}{r_a}$$

$$\phi_s^2 = \frac{r_a^2 C_x V_{s,max}}{K_{m,s} D_{e,s}}$$

$$Bi_s = \frac{\varepsilon_a r_a K_{l,s}}{D_{e,s}}$$

and $g_s [w_s (z)]$ is a dimensionless reaction rate expression, some common forms of which are summarized in Table 9.1.

The influence of intra- and extra-particle diffusional limitations can be characterized by an effectiveness factor which is defined as the total rate of reaction to that based on the bulk substrate concentration:

$$\eta = \frac{\int_0^1 (n+1)z^n g_s\{w_s(z)\}dz}{g_s\{w_s(1)\}}$$

Table 9.1. Typical dimensionless expressions for substrate uptake rate

Michaelis - Menten (M-M)	$\dfrac{w_s}{1 + \kappa_1 w_s}$
M-M with non-competitive substrate inhibition	$\dfrac{w_s}{\left(1 + \kappa_1 w_s\right)\left(1 + \kappa_2 w_s\right)}$
M-M with competitive substrate inhibition	$\dfrac{w_s}{1 + \kappa_1 w_s + \kappa_3 w_s^2}$
Extended M-M with substrate inhibition	$\infty\left(1 - \kappa_4 w_s\right)^\alpha$

$\kappa_1 = C_{s,b}/K_{m,s}$ $\qquad \kappa_2 = C_{s,b}/K_{i,s}$ $\qquad \kappa_3 = (C_{s,b})^2/K_{m,s}K_{i,s}$ $\qquad \kappa_4 = C_{s,b}/C_{s,max}$
$K_{m,s}$ is the substrate saturation constant, $K_{i,s}$ is the substrate inhibition constant and $C_{s,max}$ is the maximum tolerable substrate concentration.

For most practical cases the integral form of the effectiveness factor is equivalent to the following differential form:

$$\eta = \frac{n+1}{\phi_s^2 g_s\left[w_s(1)\right]} \frac{dw_s(1)}{dz}$$

A number of assumptions are implicit in the above formulations:
- The aggregate is isothermal
- Fickian diffusion is the only mass transfer mechanism
- The aggregate can be represented as a homogeneous phase
- A mass transfer coefficient defines external transport to the aggregate
- The aggregate is at steady state
- A single spatial variable suffices to describe the substrate concentration profile
- There is one diffusing species–the rate limiting substrate
- There is one reaction taking place
- There is no spatial dependence of the cell concentration
- There is no spatial dependence of the diffusivity

However little justification there may be in accepting that all these assumptions are valid, the above form of the diffusion/reaction equation is the most often considered in the context of immobilized living cells. With a few exceptions, e.g. a first order reaction expression, analytical solutions of this simplified diffusion/reaction equation cannot be readily obtained and recourse to algebraic approximations or numerical techniques is necessary. A large number of publications have addressed this problem in the past. It would appear, however, that most of the published work is concerned with specific aspects of the problem, e.g. particular geometries and rate expressions, or a limited range of operating conditions, e.g. $Bi \to \infty$ and small values of ϕ. In certain cases, the choice of conditions may have been restricted by the robustness of the numerical technique.

Despite its limitations, diffusion-reaction theory provides invaluable insights into the factors affecting the behavior of immobilized cell aggregates (Example 12: Ethanol Production by Immobilized Cells of *Saccharomyces uvarum*).

MISUSE OF CONVENTIONAL DIFFUSION-REACTION MODELS

It is often the case that diffusion-reaction models are being used to describe the behavior of immobilized viable cell systems without a critical appraisal of the underlying assumptions. Furthermore, more, in certain instances the actual models used are mathematically and conceptually incorrect. A first example comes from the work of Hamamci and Ryu[2] who simulated the performance of a tapered column bioreactor employed in ethanol production by linking the reactor mass balance equations with the diffusion-reaction equations which apply at the immobilized cell aggregate scale. They used the following form of the diffusion-reaction equation:

$$
D_e \left(\frac{d^2 C}{dr^2} + \frac{2}{r} \frac{dC}{dr} \right) = \frac{V_{m1} C}{C + K_{m1} + \dfrac{C^2}{K_{is}}} + \frac{V_{m2} C}{C + K_{m2} + \dfrac{C^2}{K_{is}}}
$$

while the two concentration domains were linked by:

$$
K_L a \left(C_b - C_s \right) = \eta \left(\frac{V_{m1} C_s}{C_s + K_{m1} + \dfrac{C_s^2}{K_{is}}} + \frac{V_{m2} C_s}{C_s + K_{m2} + \dfrac{C_s^2}{K_{is}}} \right) \left(1 - \frac{P_s}{K_p} \right)
$$

where C_b is the concentration of substrate in the bulk, C_s, P_s are the concentrations of substrate and product at the particle surface, V_{m1}, V_{m2} were defined as maximum specific ethanol formation rates (g ethanol/g cell/h), K_{m1}, K_{m2}, K_{is}, K_p are kinetic constants (g/L), $K_L a$ is the mass transfer coefficient (1/h) and η is the effectiveness factor of substrate uptake.

A quick glance at the above equations shows that the reaction rate expression in the right hand side of the diffusion-reaction equation is somewhat shorter (and perhaps easier to handle) than the one used in the reactor mass balance equation. More alarming, however, is the fact thatthe above equations are not only inconsistent but also quite wrong. The left hand side of the diffusion-reaction equation, which denotes the spatial change in the diffusive flux of the substrate across the particle, should be equal to the volumetric rate of substrate consumption and not to the specific rate of product formation, as implied in these equations. A multiplication factor of C_x $(1/Y_{ps})$, where C_x is the biomass concentration per particle volume and Y_{ps} is the yield of ethanol on glucose, should be used in the reaction rate expression in order to make the equation, at least, dimensionally consistent. The same applies to the reactor equation.

A second example comes from Jain et al[3] who estimated the maintenance coefficient m of immobilized *Zymomonas mobilis* cells employed in the production of ethanol from fructose by using the following equation:

$$q_s = \frac{\mu}{Y_{x/s}} + m + \frac{q_p}{Y_{p/s}}$$

where μ is the specific cell growth rate, m is the cell maintenance coefficient and q_p is the specific ethanol production rate. The above equation is quite wrong, as the first and third term in the right hand side refer to the same quantity. This equation would stand only for products indirectly linked to energy generation by the cells, which is not true of ethanol production. In addition, the authors base their calculation on the assumption that the growth rate of immobilized cells is equal to the dilution rate which, although valid in free cell culture, is totally untrue in immobilized cell fermentations.

POOR USE OF CONVENTIONAL DIFFUSION-REACTION MODELS

It remains yet to be investigated whether the assumptions underlying conventional diffusion reaction models hold true in the case of immobilized living cells. The evidence so far suggests that this may not be the case. Karel et al[4] concluded that there is a lack of understanding of the fundamental properties of immobilized cells which makes their modelling a very difficult task. A number of reasons why such conventional models may be hopelessly inadequate follow.

- Within a living cell there is a whole network of reactions taking place which involve a number of different substrates, intermediates and products. The common approach in modelling these systems is to consider a single reaction and a single diffusing species, the rate limiting substrate. This may be inadequate, especially for metabolites which are not directly associated with the generation of energy by the cells and for reactions which involve inhibitory products.

- It is quite common to consider a uniform biomass distribution throughout the immobilized cell aggregate in formulating the relevant models. However, more often than not, cells would exhibit a heterogeneous cell distribution, especially when they are grown in situ. Additionally, there are severe substrate limitations or product inhibition.

- An immobilized cell aggregate is normally described as a pseudo-homogeneous phase, i.e. carrier material, interstitial fluid and individual cells are assumed to be randomly distributed throughout the particle so that there are no significant local variations of the physicochemical properties of the aggregate due to structural effects, i.e. there is a single diffusivity value which can satisfactorily characterize the system. There is, however, ample experimental evidence suggesting that immobilized cells may grow encaged within the cavities of the particle in such a way that the aggregate can no longer be considered an assembly of cells immobilized but rather as an assembly of immobilized microcolonies. Under these conditions the assumption of a homogeneous immobilized phase becomes invalid because of the existence of two very different levels of structure: the bulk support matrix and the microcolonies which contain micropores, orders of magnitude smaller than the actual pores of the particle. Thus, locally high cell concentrations and low diffusivities can result in low local effectiveness factors.

- The dynamic characteristics of immobilized cells have been largely disregarded in relevant models. A steady state is often assumed between the various competing mechanisms such as cell growth, decay, maintenance and leakage from the support. It is often the case that this steady state is associated with a uniform biomass distribution which is in contradiction with the conditions enabling such a steady state.

Despite the pitfalls of applying conventional diffusion-reaction theory to modelling the behavior of immobilized viable cell systems, there may be some rationale in using it: new formalisms (which will be explored in later chapters) are bound to be more complex and may require knowledge of parameters which are difficult to measure. A discussion of how one can make the best use of the simplest yet most inaccurate diffusion-reaction models follows.

PROPER USE OF CONVENTIONAL DIFFUSION REACTION MODELS

There are many occasions where the use of conventional diffusion-reaction models can provide useful insights into phenomena associated with immobilized cell fermentations. Some examples of such usage are presented below.

ETHANOL PRODUCTION

Although the overall performance of immobilized cell systems which are employed in ethanol production is generally considerably better than that of free cell systems, productivity per unit mass of cells is usually lower. Several workers have attributed the observed low efficiency to internal and/or film-type mass transfer limitations. In order to investigate whether or not this is a reasonable imputation, the conventional diffusion-reaction model described above was tailored in such a way as to yield a realistic 'worst case' scenario for ethanol production. If, under the conditions prevailing in the worst case, no mass transfer limitations were predicted by the model then, by inference, either the model should be incorrect or the observed low efficiencies should be ascribed to some other factor, such as inadequate mixing or axial dispersion. The conditions comprising the 'worst case' are listed below:

Microorganism: *Zymomonas mobilis*

The most often employed organisms in the production of ethanol are the yeast *Saccharomyces cerevisiae* and the bacterium *Zymomonas mobilis*. The organism chosen for the worst case scenario is *Z. mobilis*, which not only is much more active than yeast strains (a typical value of the maximum specific uptake rate for *Z. mobilis*[5] is 11.3 h^{-1} compared to 1.4 h^{-1} tor *S. cerevisiae*[6]) but can also be packed much more efficiently due to it's geometrical characteristics; this can lead to higher cell packing densities and, hence, higher overall rates of reaction. The higher the rate of reaction, the greater is the requirement for fast diffusion of nutrients into and products out of the immobilized cell aggregate, making the *Z. mobilis* case the worst one for diffusional limitations.

Cell Density: Near the Cell Packing Density

The higher the biomass concentration, the higher the volumetric rate of reaction and hence the requirement for fast diffusion of substrates and products. The use of a high cell density may also contribute to a reduction in the value of the effective diffusivity which strengthens the 'worst case' claim.

Cell Distribution: Homogeneous

Cells are considered to be homogeneously distributed throughout the aggregate though, in practice, a spatial distribution of cells is usually observed, with the cells preferring to grow near the aggregate surface. The assumption of a uniform distribution should yield a lower effectiveness factor, since fewer cells are exposed to the higher substrate and lower product concentrations in the uniform case than in the nonuniform one. This would not, however, be the case if the cells were exhibiting a core-like distribution, being packed more densely at the center, and is also not valid when diffusivities exhibit a strong dependence on the biomass loading.

Bulk Substrate Concentration: Low

The effectiveness of an immobilized cell aggregate depends on the concentration of substrates and products in the bulk liquid.

When substrates are supplied at relatively low concentrations, cells inside the particle are likely to suffer from substrate limitation or may even be out of reach of the diffusing substrate. On the other hand, substrates supplied at inhibitory concentrations might be consumed faster inside the particle than at the surface because of the substrate concentration dropping below inhibitory levels. Therefore, using relatively low bulk substrate concentrations in the solution to the diffusion/reaction equation provides a 'worst case' assumption not only because of possible substrate limitation inside the aggregate but also due to the elimination of possibly advantageous diffusional limitations brought about by substrate inhibition.

Product Inhibition Effects: Incorporated in the Reaction Rate Expression

The influence of the bulk ethanol concentration on the effectiveness factor has not yet been addressed in a systematic way. If one were to adopt the oversimplified approach not to incorporate a product-inhibition term in the reaction rate expression, then it may well be that predicted effectiveness factors would be lower than those where a product inhibition term was included, due to the overestimation of the rate of reaction in the former case. In view of the significance of product inhibition in ethanol fermentations, such a 'worst case' is, however, unrealistic and, thus, a product-inhibition term is incorporated in the reaction rate expression. Product inhibition effects are expected to be higher towards the center of the particle due to the concentration gradient, therefore, the incorporation of a product inhibition term in the rate equation reinforces the 'worst case' claim.

Convective Transport: Neglected

The lack of a convective term for mass transfer in the diffusion/reaction equations results in lower overall resistances.

It is also implicit in the above assumptions that the rate limiting substrate is the carbon source and that other nutrients are supplied in excess. It may well be the case that inadequate provision of some nutrient which is not required by the reaction but which is essential for cell maintenance, e.g. oxygen for yeast cells, contributes to a low apparent stability and activity of the immobilized cells.

By choosing conditions which comprised a realistic 'worst case,' the effect of changing various system parameters was studied and substrate concentration profiles and effectiveness factors calculated. It was observed that over a wide range of practical conditions, particles of 1 mm diameter or less were always free from mass transfer limitations, i.e. effectiveness factor close to one (Fig. 9.1 shows a relatively severe case). This contrasts markedly, however, with the typically low reactor efficiencies reported in the literature for simple packed bed reactors.

A similar disparity between theoretical and experimental data was reported by Young and Royer[7] for growing yeast cells immobilized in an expandable poly-ethyleneimine alginate matrix; the experimental values were substantially lower than the theoretical values when the substrate saturation parameter reached 100.

It should be noted that the theoretical calculations were carried out using the value for the glucose diffusivity in a cell-free matrix. A possible explanation for the observed difference could therefore be the use of a relatively high value for the effective diffusivity. There is, however, conflicting evidence in the recent literature with regard to the effect of biomass loading on the effective diffusivity values. Hannoun and Stephanopoulos[8] found that the presence of 20% yeast cells had no effect on the value of the effective diffusivity, which is in contradiction with the data of Seki and Furusaki.[9] In order to elucidate the impact of a low diffusivity value on the effectiveness factor, a sensitivity analysis was performed in the 'worst case' above by reducing the effective diffusivity value by an order of magnitude. It was observed that the reported experimental efficiency factors can be accounted for only when a 'low' effective diffusivity is combined with low substrate/high product concentration in the bulk of the aqueous phase. For the set of conditions in case (a) of Figure 9.1, the use of a 'low' effective diffusivity reduces the effectiveness factor to 0.378 while for cases (b) and (c) the effectiveness factors are 0.713 and 0.744 respectively. A similar trend was observed by performing the sensitivity analysis for the cases reported in Table 9.2, though the values of the effectiveness factor are close to the value of one for medium/high substrate concentrations. It should also be added that in these cases it would be difficult to justify the reduction in the effective diffusivity value by one order of magnitude as biomass loadings are not very high (typically 60 g/L).

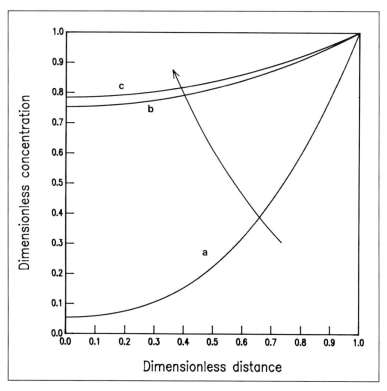

Fig. 9.1. *Substrate concentration profiles for 1 mm spherical aggregates containing 250 g/L* Zymomonas mobilis *cells exposed to glucose/ethanol concentrations of (a) 5.0, 97.5, (b) 100.0, 50.0, (c) 180.0, 10.0. The substrate concentration increases in the direction of the arrow.*

Therefore, it would appear that, unless the model is hopelessly inadequate, a major factor contributing to the low overall efficiency of the immobilized cell systems used for ethanol production is not the low effectiveness of the biocatalytic particles employed, but rather the poor design of the bioreactors with respect to the transport of nutrients and products to and from the cell aggregates (i.e. through the bulk liquid). Apart from exhibiting poor mixing characteristics due to low liquid velocities, the most commonly used type of reactor, the simple packed bed, does not provide good degassing conditions for carbon dioxide which results in dead spaces in the reactor, for channelling and even for matrix disruption. By comparison, the very few reactor efficiencies reported for relatively well mixed systems and/or systems with good degassing conditions (Table 9.3) are high.

Table 9.2. Reported packed-bed efficiencies for ethanol production by immobilized cells

Carrier	Size (mm)	Cell	Substrate	Effic.	Ref
Ca-alginate	1	*K. marxianus*	Jer. artichokes	25%	10
Ca-alginate	fibers	*Z. mobilis*	Glucose	38%	11
Ca-alginate	1	*Z. mobilis*	Glucose	52%	12

The systems in Table 9.2 were selected on the basis that a direct comparison was made between a packed reactor and the freely-suspended cell system. The inferiority of the packed bed reactor has also been demonstrated by comparing it to other reactor configurations such as a horizontal packed bed[14] or a fluidized bed reactor.[15] It might be concluded that a well mixed environment is preferable for ethanol production by immobilized cells. However, the reaction kinetics suggest ideal plug flow as being optimum. In practice, achieving plug flow conditions requires considerably better local mixing than is normally attained in simple packed beds. This and other considerations are discussed more fully in chapter 6.

OXYGEN TRANSPORT IN IMMOBILIZED CELL SYSTEMS

The transfer of oxygen from air to cells has long been recognized as one of the major bottlenecks in conventional fermentation technology. In immobilized cell systems where additional resistances to the transport of solutes are introduced, the likelihood of internal oxygen mass transfer limitation is clearly higher. The magnitude of these resistances is dependent not only on the structure of the immobilized cell aggregate but also on the biological oxygen demand of the system in question. The latter may vary widely between different systems. For example, the maximum oxygen uptake rate of respiring *Papaver somniferum* cells is reported to be 0.000432 g/g dry cells/h,[16] almost five orders of magnitude less than that of *Acetobacter aceti* cells engaged in acetic acid production (6.6 g/g dry cells/h).[17] It therefore seems almost imperative that every application should be investigated on its own merits.

The approach adopted here is to examine the relative magnitude of oxygen transport limitations in different immobilized cell

Table 9.3. Reported reactor efficiencies for ethanol production by immobilized cells

Carrier	Size (mm)	Cell	Substrate	Reactor	Effic.	Ref
κ-carrageenan	2-3	Z. mobilis	Fructose	Tower	73%	3
κ-carrageenan	2-3	Z. mobilis	Fructose	STR	90%	13
Ca-alginate	3.34	S. cerevisiae	Glucose	HPBR	113%	14

systems by looking at a wide range of applications. Underlying this analysis are the following assumptions:

- Oxygen consumption rate can be modelled using a simple Michaelis-Menten type expression.
- The particles are assumed to be spheres of 1 mm diameter; such particles are routinely made by current technology while smaller particles may be totally impracticable from an engineering viewpoint.
- Two levels of oxygen concentration in the bulk are chosen: 100% saturation as an optimistic case and 30% saturation as a more realistic one.
- The cells are homogeneously dispersed throughout the particle.
- There is no dependence of the effective diffusivity on the biomass loading. This can be justified in view of the low biomass loadings per particle used in the simulations (C_x = 1, 10 g/L)

Using the above assumptions, a number of computer simulations have been carried out using a conventional diffusion-reaction model. The following conclusions can be drawn by examining the results of the computer simulations which are presented in Figure 9.2 and Figure 9.3.

It appears that animal and plant cells should be free of oxygen mass transport limitations for the range of biomass concentrations used. In fact these systems should be free from intra-particle limitations at considerably higher biomass loadings; use of a biomass concentration for *Papaver somniferum* of 250 g/L and a value for the effective oxygen diffusivity of 1/10th of the diffusivity in water resulted in only a negligible deviation of the effectiveness factor from the value of 1.

Immobilized bacteria of *Acetobacter aceti* employed in acetic acid fermentations appear to be oxygen-limited even at a biomass concentration of 1 g/L and 100% air saturation.

The severity of oxygen mass transport limitations for respiration appears to increase in the following order: plant and animal cells, filamentous fungi, yeasts, bacteria. It is interesting to see that the above sequence also represents a classification of cells on the basis of their size.

When oxygen is incorporated in a growth- or partial-growth-associated product, the possibility of oxygen transfer limitations is increased (e.g. acetic acid, citric acid).

Some biotransformations were also included in this analysis for illustrative purposes. It can be seen that most of them fall in the diffusion-limited regime.

Only reactions with high values for the oxygen saturation constant deviate significantly from the overall trend of the effectiveness factor in relation to the max specific oxygen uptake rate.

Finally, as a note of caution, it should be stressed that although the above analysis enables an overall semi-quantitative view of oxygen transport limitations, it cannot provide accurate predictions. Oxygen-limited aggregates will typically have a heterogeneous biomass distribution which may effect significantly their reactivity.

CELLULASE PRODUCTION

Webb et al[18] reported that the specific cellulase productivity of *Trichodermal viride* cells immobilized in 6 mm diameter, spherical, stainless steel biomass support particles was more than three-fold higher than that normally obtained for freely suspended cells in continuous culture. This behavior was attributed by the authors to the low substrate concentration within a biomass support particle giving rise to an extremely low cell growth rate which in turn, induces cellulase production. In order to investigate whether or not this is a reasonable imputation, the experimental effectiveness factors infered by Webb et al[18] have been compared with theoretical ones obtained from a conventional diffusion-reaction model by assuming that the intrinsic biological kinetic parameters of the immobilized cells were those reported by Brown and Zainudeen.[19] The theoretical effectiveness factors were

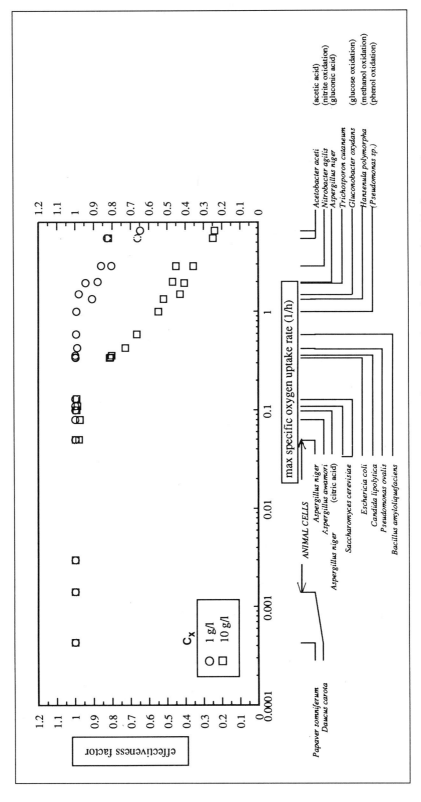

Fig. 9.2. Effectiveness factor for oxygen uptake in different biological systems for 1 mm spherical pellets and 100% air saturation.

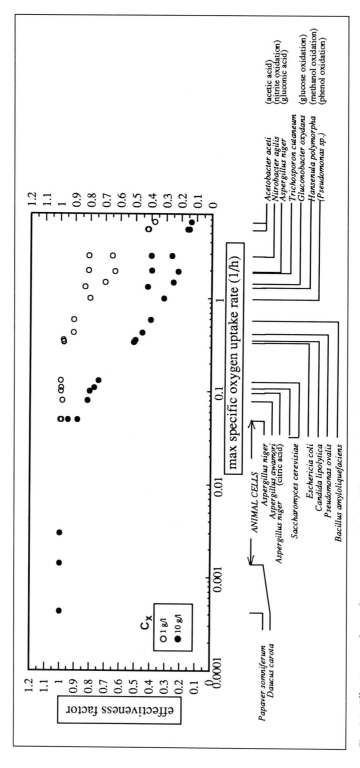

Fig. 9.3. Effectiveness factor for oxygen uptake in different biological systems for 1 mm spherical pellets and 30% air saturation.

calculated by assuming that substrate uptake followed simple Monod-type kinetics, that there was no dependence of the effective diffusivity on the biomass loading and that the spouted-bed used in the experiments behaved, essentially, as a well-mixed reactor.

The results show that although the use of a diffusion-reaction model generally provides an underestimate of the effectiveness factor, the difference is not particularly large. Nevertheless, it is worth recognizing that the theoretical predictions diverge to the opposite direction compared to that observed in the production of ethanol by immobilized cells. A possible explanation for this trend is that endogenous metabolism becomes significant under the conditions of low biomass concentration and low biomass growth rate which prevail in the immobilized cell system for cellulase production. The result of a significant endogenous metabolic rate is that the actual biomass concentration during the experiment may be smaller than the initial biomass concentration which, in turn, explains the underestimation in the effectiveness factor: the reduction in the average biomass concentration is due to part of the biomass being consumed for maintenance purposes (endogenous metabolism).

Despite the slight disparity between experimental and theoretical effectiveness factors it is worth noting that the imputation of Webb et al[18] was largely correct: an examination of the predicted substrate concentration profiles through a biomass support particle reveals that the penetration depth is less than the particle radius (Fig. 9.4); this supports the hypothesis of advantageous diffusion limitations for non-growth associated products of metabolism.

CONCLUSION

The application of an ordinary diffusion-reaction model can provide preliminary guidelines with regard to the severity of intra-particle mass transfer limitations. A number of qualitative conclusions can also be drawn for the relative importance of diffusion limitations across different systems. The application of the model to real systems has highlighted a number of potential pitfalls, listed below.

The case study for ethanol production has indicated the need to investigate the effect of high biomass loadings on the effective diffusivity and to address product inhibition effects on a systematic basis. The case study for oxygen transport has demonstrated

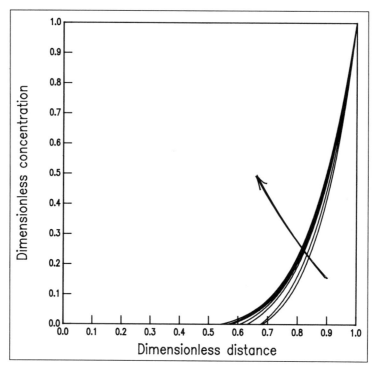

Fig. 9.4. Glucose concentration profiles for 6 mm spherical biomass support particles. The bulk substrate concentration increases in the direction of the arrow.

the need to quantify the effect of heterogeneous cell distribution on the effectiveness factor. Finally, the case study for cellulase production has highlighted the importance of considering the dynamic properties of immobilized cells when modeling their behavior.

EXAMPLE 12:
ETHANOL PRODUCTION BY IMMOBILIZED CELLS OF *SACCHAROMYCES UVARUM*

The effect of immobilization on the biological activity of respiratory-deficient yeast cells has been studied by W.C. Yu (MSc Dissertation, UMIST, 1989). In this work a petite mutant was induced from a wild-type, respiratory sufficient yeast of *Saccharomyces uvarum*. The petite mutant showed a lower biomass yield and a higher ethanol yield in comparison with the wild-type. The mutant was subsequently immobilized in Ca-alginate beads with a view towards identifying the effect of various process conditions on the efficiency of the immobilized cell particles. The latter was quantified by measuring the rate of the biological reaction in an initial rate cell and comparing it with that of freely-suspended cells.

Yu studied the effects of external and internal mass transfer resistances by varying recirculation rate, bead size, cell loading and glucose concentration. Mass transfer resistance around the beads was negligible when recirculation rate was greater than 150 ml/min. The internal mass transfer resistance was observed to increase with increasing bead size, increasing cell loading and decreasing glucose concentration. Based on the kinetics of freely suspended cells, the effectiveness factors of the immobilized yeast particles were determined with values ranging between 0.17-1.13.

Five years later, F. Heinzelmann (MSc Dissertation, UMIST, 1994) attempted an independent reappraisal of Yu's results by using a conventional diffusion-reaction model based on Michaelis-Menten type kinetics to estimate the most probable theoretical effectiveness factors in Yu's original experiments. The rationale for relying on a simple diffusion-reaction model to perform these calculations was that most of the underlying assumptions seemed to hold true. As the experiments were based on measuring initial reaction rates, the ethanol concentration in the system was quite low and it did not necessitate the use of a product inhibition term in the reaction rate expression. Substrate inhibition was also neglected as glucose concentration in the system never exceeded 5 g/L. A Biot number of 10000 was assumed in order to indicate low external mass transfer resistances. The one parameter which seemed to cause problems was the diffusivity of glucose inside the Ca-alginate beads. Yu had used fairly high cell loadings (between 76-90 g/L) and this could have had a significant effect on the value of the effective glucose diffusivity. In view of the ambiguity surrounding this parameter (which will be analyzed more fully in the next chapter), three scenarios have been proposed: an optimistic scenario where the presence of cells was assumed not to have any impact on the value of the effective diffusivity; a pessimistic scenario where the reduction in effective diffusivity was assumed to follow a Wakao-type correlation; and a most probable scenario where the reduction in effective diffusivity was assumed to follow a Maxwell-type correlation (see chapter 10).

A comparison between theoretical and experimental results can be seen in Example Figure 12.1. It is obvious that while experimental trends can certainly be predicted by the theory, actual values may differ significantly and, in a number of instances, may fall outside the area defined by the optimistic and pessimistic scenarios. If this is the case with

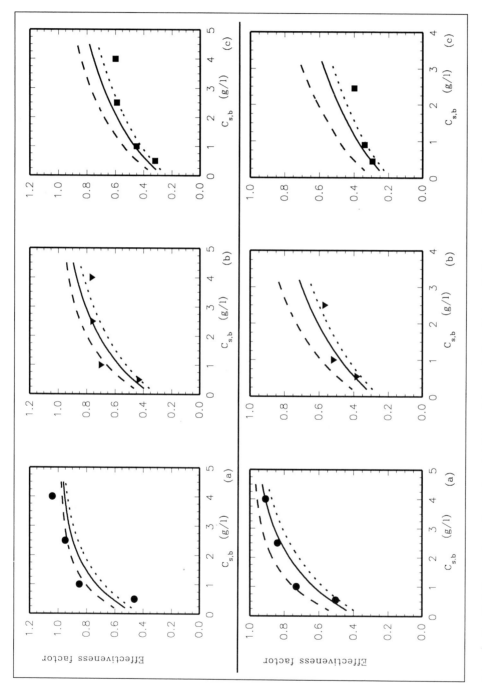

Example Fig. 12.1. Comparison of experimental data and theoretical prediction of effectiveness factors under various conditions for yeast cells immobilized in Ca-alginate beads. Bead radius: (●) 1.225 mm; (▲) 1.76 mm; (■) 2.315 mm. Correlation for effective diffusivities: (........) Wakao; (———) (Maxwell; (– – –) empty Ca-alginate beads. The top three plots are for a biomass concentration of 94 g/L; the bottom set are for a biomass concentration of 126 g/L.

experiments performed in the well-controlled environment of an initial rate cell, one can only assume that the theory needs to be revisited. This will be done briefly in chapters 10 and 11.

ACKNOWLEDGMENTS

The material for this example was adapted from the MSc Dissertation of Wan Chin Yu (UMIST, 1989) and the MSc Dissertation of Frank Heinzelmann (UMIST, 1994).

REFERENCES

1. Weisz PB. Diffusion and chemical transformation. an interdisciplinary excursion. Science 1973; 179:433-436.
2. Hamamci H, Ryu DDY. Performance of a tapered column packed-bed bioreactor for ethanol production. Biotechnol Bioeng 1987; 29:994-1002.
3. Jain VK, Toran-Diaz I, Baratti J. Preparation and characterization of immobilized growing cells of *Zymomonas mobilis* for ethanol production. Biotechnol Bioeng 1985; 27:273-279.
4. Karel SF, Libicki SB, Robertson CR. The immobilization of whole cells: engineering principles. Chem Eng Sci 1985; 40:1321-1354.
5. Lee JH, Skotnicki ML, Rogers PL. Kinetic studies on a flocculent strain of Zymomonas mobilis. Biotechnol Let 1982; 4:615-620.
6. Tyagi RD, Ghose TK. Studies on immobilized *Saccharomyces cerevisiae. I.* Analysis of continuous rapid ethanol fermentation in immobilized cell reactor. Biotechnol Bioeng 1982; 24:781-795.
7. Young JJ, Royer GP. Immobilization of growing cells and its application to the continuous ethanol fermentation process. Ann NY Acad Sci 1990; 589:271-282.
8. Hannoun BJM, Stephanopoulos G. Diffusion coefficients of glucose and ethanol in cell-free and cell-occupied Ca-alginate membranes. Biotechnol Bioeng 1986; 28:829-835.
9. Seki M, Furusaki S. Effect or intra-particle diffusion on reaction by immobilized growing yeast. J Chem Eng Japan 1985; 18:461-463.
10. Margaritis A, Bajpai P. Continuous ethanol production from Jerusalem artichoke tubers: II Use of immobilized cells of *Kluvveromyces marxianus.* Biotechnol Bioeng 1982; 94:1483-1493.

11. Grote We Lee KJ, Rogers PL. Continuous ethanol production by immobilized cells of *Zymomonas mobilis*. Biotechnol Let 1980, 2:481-486

12. Margaritis A, Bajpai PK, Wallace JB. High ethanol productivity using small Ca alginate beads of immobilized cells of *Zymomonas mobilis*. Biotechnol Let 1981; 3:613-618.

13. Jain WK, Toran-Diaz I, Baratti J. Continuous production of ethanol from fructose by immobilized growing cells of *Zymomonas mobilis*". Biotechnol Bioeng 1985; 27:613-620.

14. Chien NK, Sofer SS. Flow rate and bead size as critical parameters for immobilized yeast reactors. Enzyme Microb Technol 1985; 7:538-542.

15. Cho GH, Choi CY, Choi YD, Han MH. Continuous ethanol production by immobilized yeast in a fluidized reactor. Biotechnol Let 1981; 3:667-671.

16. Furusaki S, Nozawa T, Isohara T, Furuya T. Influence of substrate transport on the activity of immobilized *Papaver somniferum* cells. Appl Microbiol Biotechnol 1988; 29:437-4.

17. Sun Y, Furusaki S. Continuous production of acetic acid using immobilized *Acetobacter Aceti* in a three phase fluidized bed bioreactor. J Ferm Bioeng 1990; 69:102-110.

18. Webb C, Fukuda H, Atkinson B. The production of cellulase in a spouted bed fermenter using cells immobilized in biomass support particles. Biotechnol Bioeng 1986; 28:41-50.

19. Brown DE, Zainudeen MA. Growth kinetics and cellulase biosynthesis in the continuous culture of *Trichoderma viride*. Biotechnol Bioeng 1977; 19:941-957.

CHAPTER 10

MISINTERPRETATION OF EFFECTIVE DIFFUSIVITIES

Diffusion and reaction theory has never ceased to fascinate biologists and engineers. This theory is, somehow, endowed with a unique ability to provide simple explanations to a great number of enormously complex yet seemingly unrelated phenomena. It has been used to interpret all sorts of diverse phenomena, including the establishment of morphogenetic gradients in embryonic development, the size of cellular and sub-cellular components, the length of the tail of spermatozoa and the lower reactivity of enzyme crystals compared to their dissolved counterparts.

As expected, diffusion and reaction theory pervades many disciplines. In the preface of his two-volume treatise on diffusion and reaction in permeable catalysts, Aris[1] warns biologists of the futility of re-inventing the theory of diffusion and reaction:

"... Finally the term 'permeable', rather than 'porous', has been used intentionally in the hope that it will strike a sympathetic chord in the biologist's ear. It is becoming increasingly obvious that the reaction and diffusion problems in biology have much in common with those of chemical engineering. This awareness of the pervasive influence of diffusion in biology ... is seen everywhere in the literature and there is some danger of unnecessary duplication of effort..."

EFFECTIVE DIFFUSIVITY

The unnecessary duplication of effort has, apparently, been averted in the area of viable cell immobilization where diffusion-reaction formalisms are being used in an attempt to predict the biological activity of immobilized cell aggregates. However, the effectiveness of the theory in this area is doubtful. Immobilized cell aggregates have proved thus far to be particularly resistant to any sort of quantitative treatment within a diffusion-reaction framework. The thesis of this communication is that this failure can be largely attributed to misinterpretation of the concept of effective diffusivity which, for immobilized cells, has a connotation different than that for porous catalysts.

Effective diffusivity of a solute inside a particle, D_e, is the property which quantifies the resistance that the particle structure exerts on the transport of the solute. Our ability to measure and predict the value of this property is critical for the meaningful application of diffusion-reaction theory. In the area of immobilized cells, however, the problem is more fundamental. The problem is not so much one of measuring effective diffusivities but one of understanding and interpreting what we measure. A cursory glance at Figure 10.1, which summarizes the effect of particle biomass holdup on the value of effective diffusivity, will perhaps prove this point; there is considerable scatter in the experimental data for effective diffusivity values[2-22] and little agreement in the correlations proposed to fit these data.[23-27]

A FUNDAMENTAL PROBLEM

At the root of the problem is the misinterpretation of the diffusion-reaction equation which, in simple terms, states that the rate of diffusion of a solute through a particle equals its rate of consumption by the biological reaction:

$$D_e \nabla^2 c = R(c)$$

where D_e is the effective solute diffusivity, ∇ is a differential operator and R is the volumetric biological reaction rate which depends on the concentration of solute c.

The above equation, however, is so often abused that it very rarely equates like with like. For the equation to hold, the volume basis for the solute concentration c should be the same on both sides of the equation. In the case of the reaction term, this basis

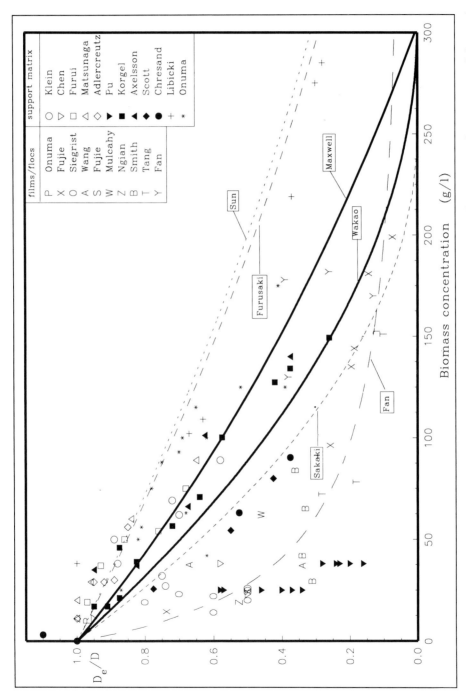

Fig. 10.1. Effect of particle biomass holdup on the relative effective diffusivity. There is considerable scatter in the experimental data and little agreement amongst the six correlations proposed to fit these data. D is the molecular diffusivity in water and D_e is the effective diffusivity in the immobilized cell particle.

is, more often than not, *the volume of the inter-cellular fluid within the particle* (Fig. 10.2). In the case of the diffusion term, however, the volume basis for the solute concentration depends on the way that D_e has been estimated. For example, if D_e has been measured in the absence of cells, then the volume basis for c should be the volume taken up by the fluid inside a cell-free particle. If D_e has been measured in the presence of cells which can be permeated by the solute, then the correct volume basis for c should be the inter-cellular fluid within the particle plus the volume of cells which is

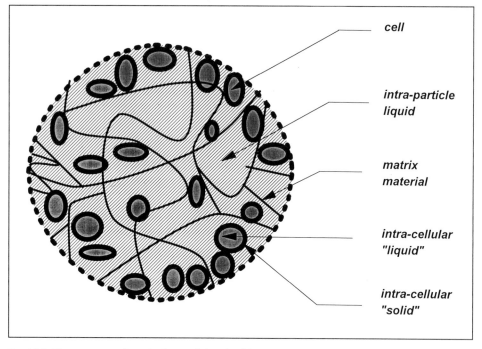

Fig. 10.2. The structure of a typical immobilized cell particle. For the theory of diffusion and reaction to hold, the experimental technique for measuring D_e should secure that the solute is traveling only within the intra-particle liquid (light-shaded area).

accessible to the solute. In both cases, it would be wrong to apply the above equation as it does not equate like with like. These are only a couple of cases which show that the concept of D_e has been abused in the literature. A detailed list and a comprehensive analysis of common errors in measuring and interpreting effective diffusivities follows (Fig. 10.3).

THE PRESENCE OF CELLS CANNOT BE IGNORED

When D_e is estimated from experiments performed using immobilization particles lacking cells, then its value is likely to be overestimated. The reason for the overestimation is that an empty immobilization carrier is bound to have a higher porosity and a lower tortuosity than a cell-occupied one. The overestimation can be quite considerable. Luong[28] estimated D_e for glucose in κ-carageenan particles containing *Zymomonas mobilis* cells by ignoring the presence of biomass. It was concluded that at a level of glucose concentration at the particle surface of 10 g/L and using particles 1 mm in diameter, the biomass concentration in the particle that would give an effectiveness factor of 0.95 could be as high as 276 g dry cells/L particle. However, even if one assumes a 'weak' decrease of D_e with cell concentration and uses a Maxwell-type relationship to describe the reduction (Fig. 10.1), then one can arrive at a value for the particle effectiveness factor of only 0.36, invalidating some of the conclusions in Luong's paper. Should one have used a 'strong' correlation for the effective diffusivity with cell concentration (e.g. Wakao-type, Fig. 10.1), the difference would have been even larger.

CELLS ARE SELECTIVELY PERMEABLE

When a non-metabolizable solute has the ability to penetrate through the cell membrane of immobilized cells then D_e is overestimated; the immobilized cells may either lack the relevant enzymatic systems to take up the solute or they may have been made inactive. Inserting a diffusivity value measured in this way in the reaction/diffusion equation, would result in overestimations of the diffusive fluxes and possible overestimates of the substrate uptake rate. Karel and Robertson[30,29] estimated D_e for glucose and oxygen in dense immobilized cell aggregates of *E. coli* based on the relative diffusivity of N_2O inside *E. coli* cells. While this may sound like a reasonable assumption, it has to be borne in mind that N_2O

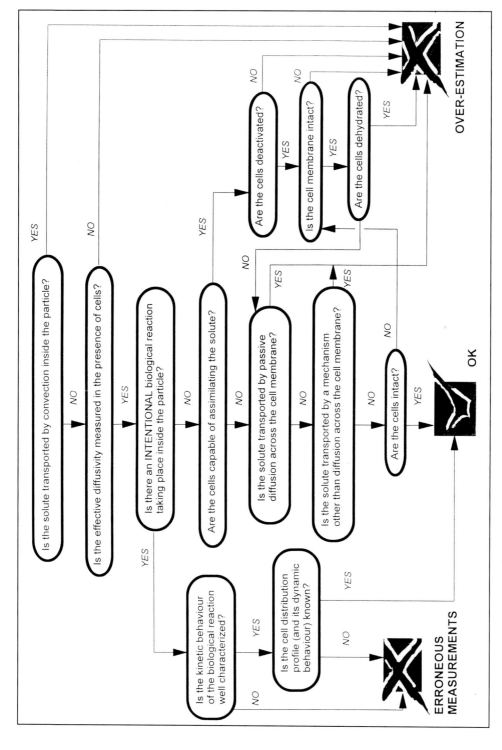

Fig. 10.3. Decision tree to assist the interpretation of experimental effective diffusivity values.

is not excluded by the cell membrane and has the ability to permeate cells.

MEASURING REACTIVE D_e CAN BE TRICKY

When D_e is measured intentionally under reactive conditions and the exact form of the rate expression is not known, then the value calculated for D_e is obviously false. A classic example is the paper by Yano et al[31] who used a zero-order rate expression in order to estimate D_e for oxygen inside mold pellets. In a re-appraisal of the original analysis by Yano and co-workers, Libicki et al[11] used a more appropriate kinetic model and obtained values that were ca. 85% larger than those calculated using zero-order kinetics. Knowing the exact form of the reaction rate expression, however, may not be enough. In situ cultivated cells are apt to be distributed non-uniformly throughout the particle thereby resulting in erroneous measurements.

CELL INACTIVATION CAN BE TRICKY

When D_e is measured intentionally by a non-reactive technique and the solute has the ability to be transported into and assimilated by non-properly inactivated immobilized cells, then the D_e estimate is inaccurate. Smith and Coakley[20] reported that while UV irradiation was ineffective in suppressing the respiration rate of activated sludge in their measurements of the effective oxygen diffusivity, addition of $HgCl_2$ proved adequate. Inactivation by heat may well be insufficient, especially when cells are heat-resistant.

When the non-reactive effective solute diffusivity is measured in the presence of 'inactivated' cells which, as a result of the 'inactivation' process, have their membranes permeabilized, then D_e is likely to be overestimated. Hannoun and Stephanopoulos[32] used 50% ethanol to kill the yeast cells in their measurements of glucose diffusivity through Ca-alginate membranes. The complete cell death was possibly accompanied by permeabilization of the cell membranes which may explain the observed insensitivity of the diffusivity value in relation to the yeast content of the membrane. Libicki et al,[11] however, treated bacteria of *E. coli* with Triton X-100 detergent and found no discernible effect on the effective diffusive permeability of N_2O in comparison with the untreated bacteria. This can be attributed to the fact that N_2O can diffuse passively through the intact cell membrane and cell permeabilization is not expected to have any appreciable effect.

When the non-reactive effective solute diffusivity is measured in the presence of 'inactivated' cells which, as a result of the 'inactivation' process are dehydrated, then the effective diffusivity value is overestimated. This is due to the fact that while dehydrated cells contain the same amount of dry cell material as intact cells, they occupy less space. Solutes excluded by the cell membrane of dehydrated cells will have more intercellular space available for transport.

CONVECTION CANNOT ALWAYS BE IGNORED

When the experimental conditions for the measurement of D_e allow the solute to be transported partially by convection inside the particle, then the value of D_e is overestimated. The overestimation in the diffusivity values can be easily explained by the fact that convection is a transfer process much faster than molecular diffusion. Convective transport often manifests itself by abnormally high diffusivity values. Miura et al[33] fitted their experimental oxygen uptake rates of mycelial pellets using a diffusion/reaction model by assuming that D_e for oxygen inside the mycelial pellets was 2-4 times higher than the value of the molecular diffusivity inside the pellets. Although Ngian and Lin[14] attributed the above values to the breaking-up of the pellets, it is more likely that the intra-pellet oxygen transport was enhanced by convection.[34] In another instance, Hannoun and Stephanopoulos[32] found that the ratio of ethanol diffusivity to glucose diffusivity in 1% Ca-alginate membranes was higher than the corresponding value for dilute water solutions and 2 and 4% alginate gels. The deviation was attributed to convective transport.

DIFFUSION AND BIOLOGICAL REACTION:
A NEW PARADIGM?

There is now considerable evidence that immobilized viable cells do not behave in the orderly fashion that would justify the application of simplified diffusion-reaction models to describe their behavior. Immobilized viable cells are rarely dispersed homogeneously throughout immobilized cell aggregates. They often form discrete inclusions, the number and size of which depends on their spatial position within the particle. In addition, immobilized biomass is not static, it is continually synthesized, degraded and, in a number of cases, physically removed from the aggregate. All the evidence points to the need for a more realistic representation of what is going on within the immobilized cell aggregate. It is rather incongruous that the intrusion of the chemical engineering literature in the area of heterogeneous catalysis may have inhibited our understanding of these systems by establishing mindsets and incorrect nomenclature. Perhaps a combination of image analysis of immobilized cell aggregates and Monte Carlo computations of their simulated structures may improve our understanding of these systems and lead to more rational bioreactor development.

REFERENCES

1. Aris R. The mathematical theory of diffusion and reaction in permeable catalysts (Vol 1: The theory of the steady state), Clarendon Press, Oxford, UK. 1975:1-2.
2. Adlercreutz P and Mattiasson B. Oxygen supply to immobilized cells. 4. Use of p-benzoquinone as an oxygen substitute. Appl Microbiol Biotechnol 1984; 20:296-302.
3. Axelsson A and Persson B. Determination of effective diffusion coefficients in calcium alginate gel plates with varying yeast cell content. Appl Biochem Biotechnol 1988; 18:231-250.
4. Chen KC and Huang CT. Effects on the growth of *Trichosporon cutaneum* in calcium alginate gel beads upon bed structure and oxygen transfer characteristics. Enzyme Microb Technol 1986; 10:284-292.
5. Chresand TJ, Dale BE, Hanson SC and Gillies RJ. A stirred bath technique for diffusivity measurements in gel matrices. Biotechnol Bioeng 1988; 32:1029-1036.
6. Fan L-S, Leyva-Ramos R, Wisecarver KD and Zehner BJ. Diffusion of Phenol through a biofilm grown on activated carbon particles in a draft-tube three-phase fluidized-bed biorector. Biotechnol Bioeng 1990; 35:279-286.
7. Fujie K, Tsukamoto T and Kubota H. Reaction kinetics of wastewater treatment with microbial film. J Ferment Technol 1979; 57:539-545.
8. Furui M and Yamashita K. Diffusion coefficients of solutes in immobilized cell catalysts. J Ferment Technol 1985; 63:167-173.
9. Klein J and Schara P. Entrapment of living microbial cells in covalent polymeric networks. II. A quantitative study on the kinetics of oxidative phenol degradation by entrapped *Candida tropicalis* cells. Appl Biochem Biotechnol 1981; 6:91-117.
10. Korgel BA, Rotem A and Monbouquette HG. Effective diffusivity of galactose in Ca-alginate gels containing immobilized *Zymomonas*

mobilis. Biotechnol Progr 1992; 8:111-117.

11. Libicki SB, Salmon PM and Robertson CR. The effective diffussive permeability of a non-reacting solute in microbial cell aggregate. Biotechnol Bioeng 1988; 32:68-85.

12. Matsunaga T, Karube J and Suzuki S. Some observations on immobilized hydrogen-producing bacteria. Biotechnol Bioeng 1980; 22:2607-2613.

13. Mulcahy LT, Shieh WK and LaMotta EJ. Experimental determination of intrinsic denitrification constants. Biotechnol Bioeng 1981; 23:2403-2406.

14. Ngian KF and Lin SH. Diffusion coefficient of oxygen in microbial aggregates. Biotechnol Bioeng 1976; 18:1623-1627.

15. Onuma M and Omura T. Mass transfer characteristics within microbial systems. Water Sci Technol 1982; 14:553 - 562.

16. Onuma M, Omura T, Umita T and Aizawa J. Diffusion coefficient and its dependency on some biological factor. Biotechnol Bioeng 1985; 27:1533-1539.

17. Pu HT and Yang RYK. Diffusion of sucrose and yohimbine in Ca-alginate gel beads with or without entrapped plant cells. Biotechnol Bioeng 1988; 32:891-896.

18. Scott CD, Woodward CA and Thompson JE. Solute diffusion in biocatalyst gel beads containing biocatalysis and other additives. Enzyme Microb Technol 1989; 11:258-263.

19. Siegrist H and Guijer W. Mass transfer mechanisms in a heterotrophic biofilm. Water Res 1985; 19:1369-1378.

20. Smith PG and Coackley P. Diffusivity, tortuosity and pore structure of activated sludge. Water Res 1984; 18:117-122.

21. Tang W-T and Fan L-S. Steady-state phenol degradation in a draft-tube, gas-liquid-solid, fluidized-bed bioreactor. AIChE Journal 1987; 33:239-249.

22. Wang PSC and Tien C. Bilayer film model for the interaction between adsorbtion and bacterial activity in granular activated carbon columns. AIChE Journal 1984; 30:786-794.

23. Furusaki S and Seki M. Effect of intra-particle mass transfer resistance on the reactivity of immobilized yeast cells. J Chem Eng Japan 1985; 18:389-393.

24. Maxwell JC. Electricity and Magnetism (2nd edn), Clarendon Press, Oxford, UK, 1892: 435-445.

25. Sakaki K, Nozawa T, Furusaki S. Effect of intra-particle diffusion in ethanol fermentation by immobilized *Zymomonas mobilis*. Biotechnol Bioeng 1988; 32:603- 605.

26. Sun Y, Furusaki S, Yamanchi A and Khimura K. Diffusivity of oxygen into carriers entrapping whole cells. Biotechnol Bioeng 1989; 34:55-58.

27. Wakao N and Smith JM. Diffusion in catalyst pellets. Chem Eng Sci 1962; 17:825- 834.

28. Luong JHT. Cell immobilization in κ-carrageenan for ethanol production. Biotechnol Bioeng 1985; 27:1652-1661.

29. Karel SF and Robertson CR. Cell mass synthesis and degradation by immobilized *Escherichia coli.* Biotechnol Bioeng 1989; 34:337-56.

30. Karel SF and Robertson CR. Autoradiographic determination of mass-transfer limitations in immobilized cell reactors. Biotechnol Bioeng 1989; 34:320-36.

31. Yano T, Kodama T and Yamada K. Fundamental studies on the aerobic fermentation. Part VIII. Oxygen transfer within a mold pellet. Agric Biol Chem 1961; 25:580-586.

32. Hannoun BJM and Stephanopoulos G. Diffusion coefficients of glucose and ethanol in cell-free and cell-occupied Ca-alginate membranes. Biotechnol Bioeng 1986; 28:829-835.

33. Miura Y, Miyamoto K, Kanamori T, Teramoto M and Ohira N. Oxygen transfer within fungal pellets. J Chem Eng Japan 1975; 8:300-304.

34. Miura Y and Miyamoto K. Oxygen transfer within fungal pellets. Biotechnol Bioeng 1977; XIX:1407-1409.

CHAPTER 11

HETEROGENEITIES

"...gradients are unlikely to command general acceptance until their biochemical basis is discovered experimentally, and this may not prove an easy task. Mathematically minded biologists could well object that any theory which has the same mathematical formalism would necessarily fit the observed patterns...it is my belief that mechanisms based on diffusion are not only plausible but probable. Nature usually has such difficulty evolving elaborate biochemical mechanisms (for example, those used in protein synthesis) that the underlying processes are often rather simple."

Francis Crick[1]

Several investigators have observed that there is a spatial variation in the biomass concentration within immobilized cell particles, especially when the cells require oxygen and have a high growth rate.

In one of the first demonstrations of biomass distribution heterogeneities within immobilized cell particles, Wada et al[2] entrapped a small number of growing yeast cells in 4 mm κ-carrageenan gel beads and examined their distribution in the gel particle after 16, 40 and 120 h of incubation in a continuous packed-bed reactor. It was observed that the cells grew near the particle surface and that the central part of the bead was empty of cells. Lee et al[3] estimated the active growth zone to be about 1 mm within Ca-alginate particles immobilizing yeast cells. Furusaki et al[4] observed that the packing density of *Saccharomyces formosensis* immobilized in 1.5 mm cubic polyacrylamide particles was larger by 25% within 70 μm from the surface than in the central region where cell concentration was uniform. Seki and Furusaki[5] immobilized *S. cerevisiae* in 2 - 3.3 mm Ca-alginate particles at an initial cell concentration of 40 g/L. The stationary phase of growth was reached with approximately 60 g/L cells in the center of the particle and 160 g/L cells at the particle surface, the cell concentration profile being described by an exponential function.

Osuga et al[6] incubated 3 mm κ-carrageenan gel beads containing a small number of *Acetobacter aceti* in a continuous fluidized bed bioreactor. Electron microscopic observation after 4 days showed that bacteria grew near the surface of the gel beads as clusters of colonies while the cell number in the particle center did not change significantly from the state just after immobilization, for the bacteria were not easily lysed. The depth of the active layer when the reactor was sparged with pure oxygen was 0.15 to 0.2 mm, twice as thick as with air.

Gosmann and Rehm[7] immobilized *S. cerevisiae* and *Aspergillus niger* in 2.3 mm Ca-alginate beads. In the case of *S. cerevisiae*, it was found that the formation of the dense layer in the surface area was a function of the initial cell concentration. Low initial cell concentration resulted in a uniform distribution of microcolonies, while high initial cell concentration resulted in the formation of a dense layer near the surface. The growth behavior of *Aspergillus* was dependent on the concentration of NH_4NO_3 in the medium: incubation in medium with 0.15 g/L NH_4NO_3 resulted in only the outer region of the cells showing microbial growth, while with 0.05 g/L NH_4NO_3 in the medium the mycelia filled the whole alginate beads homogeneously. However, re-incubation in growth medium resulted in a fur-like coat containing mycelia. A similar effect was observed by Kush and Rehm[8] who examined various regulation aspects of roquefortine production by mycelia of *Penicillium roqueforti* immobilized in 2.5 mm Ca-alginate beads; when the concentration of $NaNO_3$ in the growth media was 0.1 g/L, a homogeneous distribution of mycelia was obtained, while at 0.5 g/L a dense peripheral layer was formed after 21 days. Use of a yeast-extract-sucrose medium, instead of a synthetic one, also created heterogeneity in the distribution of cells. El-Sayed and Rehm[9] examined the morphology of two *Penicillium chrysogenum* strains (ATCC 12690 and S1) immobilized in Ca-alginate beads. After 96 h, strain S1 had grown mainly within a 0.25 mm subsurface layer where it continued to grow for 12 days, at which time cells near the particle center had begun to lyse. Strain ATCC 12690 initially filled the particle homogeneously but after 96 h it had formed a surface coat.

Mitani et al[10] used forced substrate supply in order to minimize the diffusional resistances in a system involving a 14 mm thick κ-carrageenan gel layer. It was observed that after 24 h of incubation the cell concentration started developing a profile, which at 250 h reached a steady state, with 1.7×10^{10} cells/mL near the surface and 4×10^9 cells/mL at the bottom of the gel layer. The results were also analyzed by a model in which the growth rate of the cells was matched by the rate of inactivation due to ethanol. Shinmyo et al[11] immobilized *Bacillus amyloliquefaciens* cells in κ-carrageenan gel beads in order to produce α-amylase. After the population reached the maximum value, cells in the gel, were restricted to a layer of 50 μm thickness from the surface of the gel, suggesting that O_2 diffusion was the growth limiting factor. Similar observations were made by Burril et al[12]

Although the phenomenon of heterogeneous biomass distribution has already been demonstrated by several authors, no systematic theoretical study has thus far been presented so far to quantify the effect of the biomass gradient on the reactivity of immobilized cell aggregates. Such a study is undertaken below where the implications of a biomass gradient inside an immobilized cell particle are analyzed for different reaction-rate expressions and different forms of biomass distribution.

THEORETICAL DEVELOPMENT

The effect of heterogeneous cell distribution on the reactivity of immobilized cell particles can be accounted for by introducing a cell distribution function $f(z)$ in the reaction-rate term of the diffusion-reaction equation. The differential mass balance for a species j, which is assumed to be involved in an energy generating reaction for the immobilized cells, may then be written in the following dimensionless form:

$$\left(\pm \delta_i \,/\, \sigma_i\right)\nabla^2 w_i = \phi_s^2 f(z) g_s\!\left[w(z)\right] \qquad (1)$$

where $f(z) = C_x/C_{x,ave}$ is the biomass distribution function which satisfies the condition:

$$\int_0^1 (n+1)z^n f(z)dz = 1 \qquad (2)$$

All other symbols are as defined in chapter 9.

As can be easily deduced from the introduction to this chapter, a multitude of cell density profiles inside immobilized cell aggregates have been reported in the literature, ranging from the uniform distribution of cells throughout the particle to their confinement within a thin layer at the aggregate surface. Although the qualitative features of most of these profiles are similar, i.e. a high relative cell concentration near the particle surface which drops towards the particle center, their exact form is very much system-specific and depends, mainly, on the severity of the mass transfer limitations inside the aggregate and the dynamic characteristics of the immobilized cells.

As there appears to be no single distribution function which would suffice to simulate all possible spatial profiles of immobilized living cells, the approach adopted here was to investigate the effect of a number of different distribution functions on the reactivity of the immobilized cell aggregate. These functions include:

The step biomass profile

$$f(z) = \alpha_1, z \geq z_c; f(z) = 0, z < z_c \qquad (3)$$

where z_c is the depth of the inert core. The constant α_1 is determined by applying the condition (2) above, i.e.

$$\alpha_1 = \frac{1}{1 - z_c^{n+1}}$$

The linear biomass profile

$$f(z) = \alpha_2(z - z_c), z \geq z_c \text{ and}$$

$$f(z) = 0, z < z_c \tag{4}$$

Similarly the constant α_2 is evaluated to be:

$$\alpha_2 = \frac{n + 2}{z_c^{n+2} - (n+2)z_c + n + 1}$$

The exponential biomass profile

$$f(z) = \alpha_3 z^p \tag{5}$$

$$\int_0^1 (n+1)z^n f(z)dz = 1 \Rightarrow \alpha_3 = 1 + \frac{p}{n+1}$$

The hyperbolic-tangent biomass profile

$$f(z) = \alpha_5(1 + \tanh(\alpha_6(z - \alpha_7))) \tag{6}$$

where α_6 determines the 'steepness' of the profile and α_7 determines the point where the concavity of the curve changes sign.

Unfortunately an analytical expression for α_5 can only be obtained for slab geometry:

$$\alpha_5 = \left\{ 1 + \left(1 / \alpha_6 \right) \log \frac{\cosh\left[\alpha_6 \left(1 - \alpha_7\right)\right]}{\cosh\left(-\alpha_6 \alpha_7\right)} \right\}^{-1}$$

Effectiveness factor

The introduction of the function $f(z)$ into the right hand-side of the diffusion-reaction equation means that the integral form of the effectiveness factor of an immobilized cell aggregate with heterogeneous biomass distribution is now given by the following modified formula:

$$\eta = \frac{\int_0^1 (n+1) z^n f(z) g_s\left[w(z)\right] dz}{g_s\left[w(1)\right]} \qquad (7)$$

The differential form should be identical to that of a particle with homogeneous biocatalyst distribution.

HYPERBOLIC-TYPE KINETICS (MICHAELIS-MENTEN)

The effect of the spatial distribution of the immobilized biomass on the effectiveness factor of an immobilized cell particle was investigated for Michaelis-Menten (M-M) type kinetics, for a number of different permutations of particle geometry n (Equation 2) and cell distribution function $f(z)$ (Equation 1), over a range of parametric values of the Thiele modulus ϕ_s and of the dimensionless substrate concentration in the bulk of the aqueous phase, κ_1 (for definitions see chapter 9).

The degree of cell heterogeneity was quantified by the introduction of a new parameter, $C_{x,s}/C_{x,ave}$, i.e. the ratio of the biomass concentration at the aggregate surface, $C_{x,s}$, to the average biomass concentration, $C_{x,ave}$. This parameter was introduced in anticipation of the fact that in the regime of strict diffusion control the biological reaction is localized in a thin layer near the particle surface which, in turn, nullifies the need to know the detailed profile throughout the particle. By using this quantity, the asymptotic behavior of the effectiveness factor for large values of ϕ_s becomes independent of the biomass profile and depends only on the biomass concentration at the surface. Another important property of the ratio $C_{x,s}/C_{x,ave}$ is that it satisfies the relationships:

$C_{x,s}/C_{x,ave}$ is most likely less than $C_{x,max}/C_{x,ave}$

$C_{x,s}/C_{x,ave}$ is almost certainly less than $\rho_b/C_{x,ave}$

where $C_{x,max}$ is the maximum reported immobilized local biomass concentration in relevant systems and ρ_b is the biomass dry density. By using this simple relationship an order-of-magnitude estimate can be made on the relative increase in the effectiveness factor of a heterogeneous cell aggregate when the average value of the biomass concentration is known but the actual details of the biomass distribution are not available.

An examination of Figure 11.1 reveals that the quantity $C_{x,s}/C_{x,ave}$ describes adequately the effect of cell distribution on the effectiveness factor, at least at the semi-quantitative level. Except for the very general conclusion that the higher the value of $C_{x,s}/C_{x,ave}$ the higher the relative increase in the effectiveness factor η, a number of other inferences can also be made:

- At low values of the Thiele modulus, ϕ_s, the relative increase in the effectiveness factor, η, is rather small and reaches a plateau. This can be explained by the fact that the effectiveness factor of the heterogeneous aggregate for M-M kinetics is bound by the value of 1, which does not leave significant room for improvement.
- At intermediate values of ϕ_s the relative increase in η is influenced to a large extent by the relative substrate concentration in the bulk of the aqueous phase κ_1 and the details of the biomass profile. In general, the lower the value of κ_1 and the steeper the biomass profile, the higher the relative increase. The impact of $C_{x,s}/C_{x,ave}$ on

η is illustrated in Figure 11.1 for ϕ_s = 5, 10; the room for improvement is clearly larger when the immobilized cell aggregate is 'more' diffusion limited, e.g. the increase is higher at a value of κ_1 = 0.2 than at a value of κ_1 = 2.

- At large values of ϕ_s, the relative increase in the effectiveness factor becomes independent of the values of ϕ_s and κ_1 and appears to be proportional to $[C_{x,s}/C_{x,ave}]^{0.5}$ (from the slope of the log-log graphs). *This represents the upper limit for the increase in the effectiveness factor.*

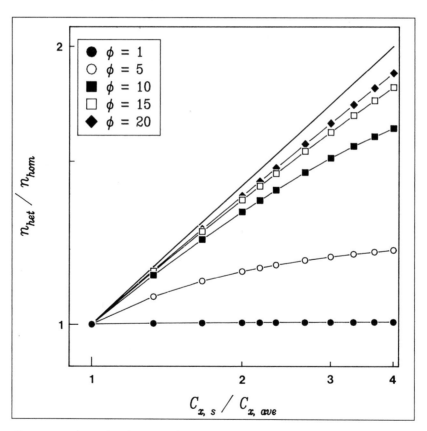

Fig. 11.1. Relationship between the effectiveness factor and the relative biomass concentration at the aggregate surface for spherical geometry, Michaelis-Menten kinetics and exponential-type cell distribution function. $C_{s,b}/K_{m,s}$ = 2.0. $C_{s,b}$ is the substrate concentration in the bulk liquid, $K_{m,s}$ is the substrate saturation constant in Michaelis-Menten kinetics, η_{het} is the effectiveness factor of the heterogeneous aggregate and η_{hom} is the effectiveness factor of the equivalent homogeneous aggregate.

The effect of cell distribution on the intra-particle substrate concentration profile is illustrated in Figure 11.2. As the value of $C_{x,s}/C_{x,ave}$ increases, the relative biomass concentration increases in the region near the particle surface and decreases in the vicinity of the particle center. As a result, the rate of reaction becomes higher near the particle surface and lower towards the center, the implication of which is that the substrate concentration profile becomes steeper near the surface and flatter near the center.

SUBSTRATE-INHIBITED REACTIONS

Although not extensively verified, it is fairly well accepted that cell immobilization can lead to advantageous diffusion limitations for substrate inhibited reactions. This is–largely–a theoretical find-

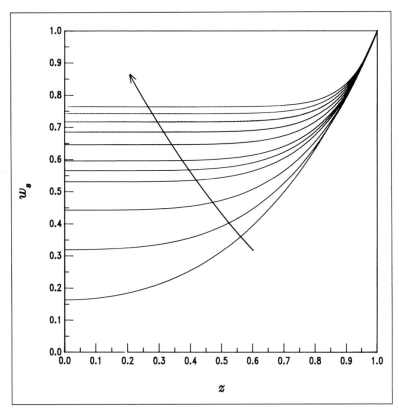

Fig. 11.2. Effect of cell distribution on intra-pellet substrate concentration profiles for spherical geometry, Michaelis-Menten type kinetics, Thiele modulus: $\phi = 5$ and $C_{s,b}/K_{m,s} = 2.0$. Cell heterogeneity increases in the direction of the arrow; w_s is the dimensionless substrate concentration and z is the dimensionless distance through the particle.

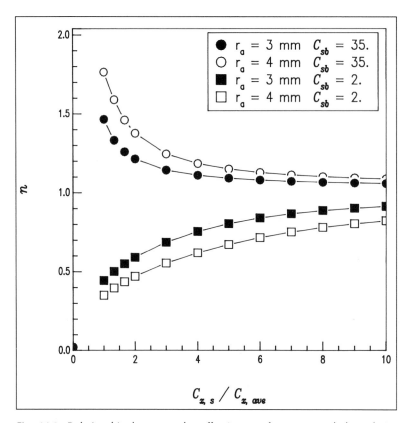

Fig. 11.3. Relationship between the effectiveness factor, η, and the relative biomass concentration at the particle surface, $C_{x,s}/C_{x\,ave}$, for a substrate-inhibited reaction, spherical geometry and exponential-type cell distribution function, at different particle radii, r_a, and bulk substrate concentrations, C_{sb}.

ing and is based on the assumption of uniform cell concentration throughout the particle. The reason for the increased effectiveness of the immobilized cells is that cells within the particle are exposed to substrate concentrations which are lower than the bulk substrate concentration and this, in turn, may result in higher intra-particle rates of reaction and higher effectiveness factors. In such cases effectiveness factors greater than one can be obtained. An example can be seen in Figure 11.3.

The reaction expression in this particular example was chosen so that it conforms to the recently proposed general reaction scheme for substrate-inhibited reactions:[13]

$$R_s = \frac{V_{s\,\max} C_s}{K_{m,s} + C_s} \left(1 - \frac{C_s}{C_{s,\max}}\right)^b$$

Parametric values for the above equation were chosen from the system *Candida utilis* /sodium acetate ($V_{s,max}$ = 0.41 1/h, $K_{m,s}$ = 0.896 g/L, $C_{s,max}$ = 39.22 g/L, b = 0.446). This reaction exhibits a maximum at C_s = 7.542 g/L. It can easily be seen that for the case of homogeneous cell distribution ($C_{x,s}/C_{x,ave}$ = 1) and for sufficiently large (35 g/L) bulk substrate concentrations, the effectiveness factor exceeds the value of one. In addition, and possibly counter-intuitively, the bigger the particle size, the higher the particle effectiveness.

Concentrating the biomass near the particle surface tends, however, to counter-balance the beneficial effect of immobilization on substrate inhibited reactions. As can be deduced from Figure 11.3, the higher the ratio $C_{x,s}/C_{x,ave}$, the lower the value of the effectiveness factor which also seems to converge for large $C_{x,s}/C_{x,ave}$ to the value of one. This behavior can be explained by the fact that concentrating the cells near the surface means that they are taken away from areas of low substrate concentration/high reaction rate with obvious implications for the particle effectiveness. At the extreme, when the biomass is localized at the surface there is no beneficial effect of cell immobilization.

The effect of the non-uniform biomass distribution in substrate inhibited reactions is a close function of the substrate concentration in the bulk. It appears that when the substrate concentration in the bulk has an advantageous effect on the effectiveness factor of a uniformly immobilized biocatalyst then a distributed cell growth will tend to counter-balance this effect. Similarly, when the substrate concentration in the bulk has a disadvantageous effect on the effectiveness factor of a uniformly immobilized biocatalyst then a distributed cell growth will tend to increase the effectiveness factor (Fig. 11.3 for $C_{s,b}$ = 2.0 g/L).

PRODUCT-INHIBITED REACTIONS

For particles having homogeneous biomass distribution, it is expected that product inhibition would be amplified by immobilization due to the establishment of an unfavorable intra-particle product gradient; cells within the particle would be exposed to product concentrations higher than the bulk product concentration and this would reduce their activity. Shifting the biomass towards the particle surface would take the cells away from areas of high product concentration/low reaction rate and, thus, it would increase their effectiveness. This is illustrated in Figure 11.4; the higher the cell heterogeneity, the higher the effectiveness factor. It should be noted however, that appreciable differences exist only when the effective product diffusivity is relatively low. This is a reflection of the fact that the lower the value of the effective product diffusivity, the larger the product concentration gradient, and the stronger the product inhibition effect, the bigger the room for improvement by employing a non-uniform biomass profile.

COMPARISON WITH LITERATURE DATA

Gosmann and Rehm[7] determined an effectiveness factor for oxygen uptake by *S. cerevisiae* and *A. niger* immobilized in Ca-alginate beads. They reported that the cells formed a dense layer

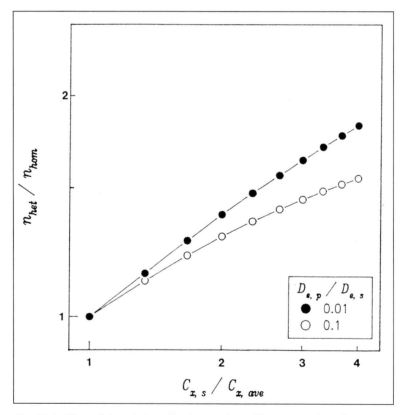

Fig. 11.4. Effect of the relative effective product diffusivity on the effectiveness factor of a spherical particle for linear product-inhibition and exponential-type cell distribution function. $D_{e,s}$ and $D_{e,p}$ are the effective diffusivities of substrate and product respectively, $C_{x,s}/C_{x,ave}$ is the relative biomass concentration at the particle surface, η_{het} is the effectiveness factor of the heterogeneous aggregate and η_{hom} is the effectiveness factor of the equivalent homogeneous aggregate.

in the outer region of the gel beads and they also examined the effect of cell concentration on the effectiveness factor. Their light microscope photographs of sliced gel beads containing a 'high' yeast cell concentration show a dark area near the particle surface which appears to have a depth between 0.1-0.2 of the particle radius, while the interior of the particle appears to be only sparsely populated with yeast cells. Assuming that the contribution of the inner cells to the reactivity of the particle is relatively small—due to their low concentration and efficiency—one can approximate the cell distribution by a step function with a dimensionless core radius z_c between 0.8-0.9 which yields a cell concentration at the particle surface which is between 2.0-3.7 times the average cell concentration (parameter α_1 in Equation 3). For a diffusion-limited particle this would correspond to an increase in the effectiveness factor of the non-homogeneous particle over the homogeneous one by a factor between 1.4-1.9. A close examination of their experimental effectiveness factors reveals that in the region of 'high' yeast concentrations the effectiveness factor of the heterogeneous aggregate is consistently higher than that of the homogeneous particle by a factor between 1.4 - 2.0 which is in excellent agreement with the present theory.

Shinmyo et al[11] attributed the growth of *B. amyloliquefaciens* in a thin sub-surface layer (~ 50 μm) inside 3 mm κ-carrageenan particles to the limited supply of oxygen brought about by diffusional limitations. In order to investigate whether or not this is a reasonable imputation, Kurosawa et al[14] performed a mathematical simulation of the oxygen uptake rate of immobilized *B. amyloliquefaciens*, using a standard diffusion-reaction model and incorporating the kinetic parameters for respiration as reported by Shinmyo et al[11] The biomass was assumed to be uniformly distributed at the cell density of the bacterial surface layer which was estimated to be about ten-fold the average biomass holdup per particle.

The results of the simulation suggest that the penetration depth of oxygen in 3 mm particles is between 60 μm and 120 μm for dissolved oxygen levels in the range [0.0625 - 0.313 mM]. This is in good agreement with the experimental results and also hints that the (non-reported) dissolved oxygen concentration from the experiments of Shinmyo[11] is most likely to have been close to the lower end of the above range of oxygen concentrations, especially so at long cultivation times. The effectiveness factor for oxygen uptake for 3 mm particles and dissolved oxygen concentration of 0.0625 mM was also determined and found to be 0.08. This is in obvious disagreement with the results of Shinmyo et al[11] who found the specific respiratory activity of the immobilized cells to be 0.57 of their free counterparts.

It seems that the approach of Kurosawa et al[14] although valid for the determination of the oxygen penetration depth, produces rather large underestimates of the effectiveness factor. Using the model described above with $n = 3$, $\phi = 54.18$, $\kappa_1 = 12.5$, $C_{x,s}/C_{x,ave} = 10.44$ the value of the effectiveness factor is estimated to be 0.7, which is much closer to the experimental results. The slight over-estimate may be attributed to the uncertainty over the value of the dissolved oxygen concentration as well as to possible non-ideal reactor behavior.

A similar analysis can be performed on the data of Seki and Furusaki[5] on the reactivity of immobilized growing yeast. The ratio $C_{x,s}/C_{x,ave}$ in their experiments was reported to be $165.0/95.2 = 1.73$ which would allow for a potential increase in the value of η of the growing immobilized cells over the non-growing ones by only 1.32. A comparison of the theoretical curve for uniform cell density after cell growth with the experimental reaction rates of the growing cells indicates a small increase in the effectiveness factor for the growing-cell system which is in good agreement with the present theory.

Caution should be applied in interpreting apparent kinetic data from immobilized cell systems, especially so when the cells occupy a significant volume of the aggregate and/or when they are heterogeneously dispersed, as classical theories may not always be applicable. Chang and Moo-Young[15] analyzed various literature data on the oxygen uptake of immobilized cell aggregates with the help of a model incorporating zero-order kinetics and homogeneous cell distribution. Their analysis may well have been invalid in certain cases. For example, it was claimed that the data of Gosmann and Rehm[16] on the oxygen uptake rate of immobilized *Aspergillus niger* indicate a condition of strict-film diffusion control because the specific oxygen uptake rate was proportional to the inverse of the biomass concentration; it was claimed that if it was intra-particle diffusion controlled it should be proportional to the inverse of the square root of the biomass concentration.

It can be easily seen that the imputation of Chang and Moo-Young would be correct only in cases were a comparison is made between systems in which a change in the biomass concentration at the particle surface is accompanied by a concurrent, proportional change in the average biomass concentration throughout the particle, and this is not necessarily the case. For example, at long cultivation times the biomass at the surface may have reached its ceiling concentration and be fairly constant with time, while the biomass away from the surface may actually be depleted as a result of endogenous metabolism and maintenance. In such a case, the effectiveness factor would be proportional to the inverse concentration and the rate limiting step can be mistaken to be the mass transfer from the bulk of the aqueous phase to the particle surface.

FURTHER HETEROGENEITIES

The analysis presented above, although an improvement on the diffusion-reaction models used to describe the behavior of immobilized viable cell systems, is a long way from being a complete representation of what is going on within an immobilized cell aggregate. For example, there is considerable evidence that immobilized viable cells often grow within the aggregate in the form of micro-colonies. Furthermore, viable cells are dynamic systems with the immobilized cell mass being continuously synthesized, degraded and removed from the aggregate. These properties of immobilized cells have been captured in a general mathematical model developed by the authors which is available from their laboratory. Two applications of this model can be found in the examples.

EXAMPLE 13:
THE EFFECT OF INTRA-PARTICLE COLONIES
ON THE EFFECTIVENESS FACTOR

There is considerable evidence that immobilized viable cells often form discrete inclusions within the carrier matrix; these inclusions are developed by a large number of closely-packed cells within the cavities of the matrix. The local biomass concentration within these inclusions can be very high, even higher than the maximum possible packing density; the compression inside the cavities can be so high that it may lead to reduced cell volumes and/or fusion of individual colonies (see chapter 3).

The resulting aggregate structure is reminiscent of a catalyst particle with a bi-dispersed porosity; the inter-cellular space within the colonies constitutes the micropores and the inter-colony space comprises the macropores. The ratio of the average pore sizes of the two pore types may well be more than one order-of-magnitude, especially in the case of an aggregate entrapping a small number of densely-populated microcolonies. This has obvious implications for the value of the effective diffusivity which can no longer be considered to be the same in the phase of the macropores and in the phase of the micropores.

A combination of a significant colony size and of a small intra-colony effective diffusivity may then lead to a value of the local effectiveness factor (at the colony level) which is less than unity. In this case the analysis of diffusion and reaction interactions using the single-phase model of a mono-dispersed catalyst becomes invalid.

The size of the individual colonies can be easily estimated by examining published electron micrographs of immobilized cell aggregates containing viable cells. There is a significant variation in their size, ranging typically from a few microns to a few hundred microns. The value of the intra-colony diffusivity, however, cannot as yet be accurately determined.

The uncertainty over the value of the effective diffusivity inside the densely populated microcolonies makes the validation of any bi-phasic diffusion/reaction model a very difficult task and any comparisons with experimental data a rather ineffectual exercise. The appeal of using such a model is further lessened by the need to determine the biomass concentration within individual colonies, the need to determine the intra-particle distribution of colonies and the possibility of metabolic disturbances induced by the close cell proximity inside the colonies. Nevertheless, it is quite important that a such a model is developed in order to provide some insight into the phenomenon and the implications of intra-particle colony formation. Such a model has been developed by the authors and some results obtained through its use are given in this example.

THEORETICAL BASIS

The model developed to account for the existence of micro-colonies within immobilized viable cell particles was based on the following assumptions:

- The aggregate has a bi-dispersed porosity: an intra-colony porosity (microporous fraction) and an inter-colony porosity (macroporous fraction).
- The effective macropore diffusivity of species i, is mainly a function of the macropore fraction and the effective micropore

diffusivity of species *i*, is mainly a function of the micropore fraction.

- A single reaction suffices to describe the biological activity of the cells.
- The reaction follows some form of Michaelis-Menten type kinetics.
- There is a spatial dependence of the reaction rate due to non-negligible local mass transfer limitations and due to an heterogeneous colony-distribution throughout the particle.
- The colonies are homogeneously populated by cells which exhibit their maximum possible packing density.

These were used in formulating sets of differential equations which were solved simultaneously, with their associated boundary conditions, using the orthogonal collocation technique. The technique was validated by checking the numerical results against the analytical solution for first order kinetics, spherical geometry and homogeneous distribution of microcolonies.

RESULTS AND DISCUSSION

A number of computer runs were performed for different values of the parameters $\phi_{s,A}$ (aggregate Thiele modulus), $\phi_{s,C}$ (colony Thiele modulus) and κ_1 (dimensionless bulk substrate concentration $-C_{s,b}/K_{m,s}$) for simple Michaelis-Menten kinetics with one diffusing species. The results were compared with the effectiveness factor of a mono-dispersed aggregate having a Thiele modulus equal to $\phi_{s,A}$ and are shown in Example Figure 13.1. A number of interesting points emerge from this analysis.

Effect of the Colony Thiele Modulus

The following two heuristics arise from the results of the computational experiments presented in Example Figure 13.1:

- When a mono-dispersed immobilized cell aggregate with uniform biomass distribution is not diffusion limited then a bi-dispersed aggregate having the same amount of immobilized biomass in the form of microcolonies can become diffusion-limited only when the colony-Thiele modulus is equal to or higher than the aggregate-Thiele modulus. This is demonstrated in Example Figure 13.1 for $\kappa_1 = 100$ and $\phi_{s,A} = 10$ or 20.
- When a mono-dispersed immobilized cell aggregate with uniform biomass distribution is diffusion limited then a bi-dispersed aggregate having the same amount of immobilized biomass in the form of microcolonies may have an effectiveness factor lower than that of the mono-dispersed aggregate at values of the

colony Thiele modulus which are smaller than the particle Thiele modulus. This is demonstrated in Example Figure 13.1 for $\kappa_1 = 0.1$ and $\phi_{s,A} = 10$ or 20.

It is quite clear that conventional diffusion-reaction models may be significantly in error, especially so in the regime of strong diffusion limitations and for relatively high values of the local Thiele modulus.

Effect of the Substrate Concentration in the Bulk of the Aqueous Phase

An important feature evident in Example Figure 13.1 is that when the bulk substrate concentration is much lower than the value of the saturation constant in the Michaelis-Menten equation ($\kappa_1 = 0.1$) then local mass transfer effects at the colony level have a notable impact on

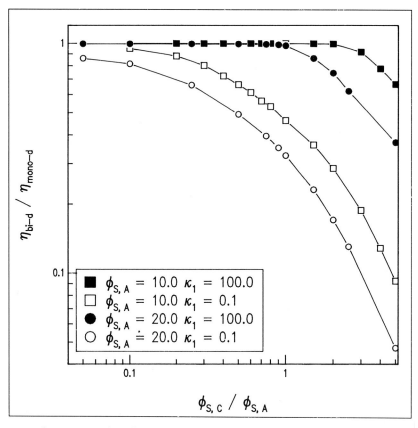

Example Fig. 13.1. The effect of the relative colony Thiele modulus, $\phi_{s,c}/\phi_{s,A}$ on the effectiveness factor of a bi-dispersed aggregate for different values of the particle Thiele modulus $\phi_{s,A}$ and the dimensionless bulk substrate concentration κ_1. (η_{mono-d} is the effectiveness factor of an immobilized cell aggregate containing homogeneously dispersed cells.)

the effectiveness factor of the aggregate, even at low values of the colony Thiele modulus. This behavior can be attributed to the fact that at such low concentrations, colonies will be consuming substrate at a linear rate with respect to the substrate concentration which, in turn, represents the lower bound of the effectiveness factor for saturation kinetics. The opposite is true for high bulk substrate concentration ($\kappa_1 = 100$); the order of the reaction kinetics with respect to substrate concentration will be, generally, less than one, which represents a favorable situation in terms of diffusion/reaction interactions.

Effect of Colony Distribution

When colonies are distributed non-uniformly within the aggregate then it is expected that the aggregate effectiveness factor will increase, since more colonies will be exposed to the higher substrate concentrations near the particle surface.

CONCLUSION

It is evident that clustering of cells in the form of colonies may have a significant impact on the reactivity of an immobilized cell aggregate, especially when the colony Thiele modulus is of the same (or higher) order of magnitude as the particle Thiele modulus. This, however, is a theoretical finding and its relevance to real systems needs to be investigated.

EXAMPLE 14:
THE DYNAMIC BEHAVIOR OF IMMOBILIZED CELLS

One of the most important features of immobilized aggregates which contain metabolically-intact cells is their dynamic nature, i.e. the immobilized cell mass is continually synthesized, degraded and, in a number of cases, physically removed from the aggregate. Although the dynamic nature of metabolically-intact cells is well acknowledged in theoretical models of traditional freely-suspended cell fermentations, it is very rarely, if ever, considered in the relevant models of immobilized cell systems. In immobilized cell systems, however, some very interesting modelling challenges exist due to the introduction of spatial heterogeneity in the intra-particle cell growth rate brought about by diffusion limitations.

A steady state model which deals with the effect of cell heterogeneity on the effectiveness factor is described in the main text of chapter 11. The aim of this example is to address some of the issues pertaining to the dynamic nature of immobilized metabolically-intact cells and to establish whether or not the observed cell distribution patterns can be explained by the theory of diffusion and reaction.

THEORETICAL BASIS

The model developed for steady state conditions was extended using the following assumptions:

- There is no spatial variation in the initial biomass concentration.
- Bulk substrate and product concentrations remain constant, i.e. the limiting nutrient is supplied continuously or/and the reactor is operated continuously. This assumption is somewhat restrictive in some immobilized cell systems. In practice, however, oxygen which is normally supplied continuously irrespective of the mode of reactor operation is often the rate limiting substrate.
- The time constant for a change in the cell concentration is much higher than the time constant for a change in the substrate concentration profile and, thus, a pseudo-steady state can be assumed in the formulation of the diffusion-reaction equations. This is because cell concentrations change with a time-constant equal to the inverse growth rate (typically of the order of a few hours) whereas the time constant for substrate diffusion is much smaller (typically of the order of a few seconds).
- Cell mass degradation is modeled with first-order kinetics.
- Any changes in the volume of the biotic phase due to cell growth/decay are accounted for in the expression for the effective diffusivity value. An 'average' value for the effective diffusivity is assumed which is a function of the average biomass holdup in the particle. The average biomass concentration per particle can be easily calculated by integrating the local biomass concentration across the whole particle using Gaussian quadrature formulas.
- There is no leakage of cells from the support. Nevertheless it is assumed that cells are unable to grow beyond a certain level of biomass concentration which is dictated, mainly, by the porosity of the support.
- There is a simple linear relationship between substrate uptake and cell growth (Y_{xs} is constant).

These were used in formulating sets of differential equations which were solved simultaneously, with their associated boundary conditions, using the orthogonal collocation technique.

RESULTS AND DISCUSSION

The model, which can be applied to any set of conditions and reaction types, was used to generate results for the dynamic behavior of an immobilized cell aggregate exposed to a constant concentration of an inhibiting substrate. These are illustrated in Example Figure 14.1 and analyzed below.

At the start of the growth period the intra-particle substrate concentration profile is such that cells located near the particle center have the highest growth rate due to the low local level of substrate concentration

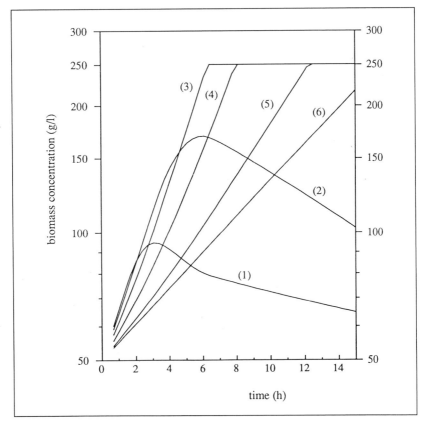

Example Fig. 14.1. Evolution of the local biomass concentration at five interior collocation points: (1) 0.2695; (2) 0.519; (3) 0.7301; (4) 0.887; (5) 0.9782 and at the particle surface (6) 1.0 for substrate inhibited kinetics. Assumed parametric values in the dynamic stimulations: maximum specific substrate uptake rate, $V_{s,max} = 0.41$ 1/h; substrate saturation constant, $K_{m,s} = 0.896$ g/L; death rate constant, $K_d = 0.1$ 1/h; maximum tolerable substrate concentration, $C_{s,max} = 39.22$ g/L; bulk substrate concentration, $C_{s,b} = 35.0$ g/L; exponent in substrate inhibition function, $b = 0.446$; effective substrate diffusivity, $D_s = 0.0004$ cm²/min; aggregate radius, $r_a = 2$ mm and initial biomass concentration in the aggregate, $C_{x,o} = 50.0$ g/L.

while cells immobilized in the outer region of the particle are growing at a lower yet finite rate. The preferential growth of cells away from the particle surface results in an initial decrease of the parameter $C_{x,s}/C_{x,ave}$ with time. As a result of the continual cell growth, the substrate concentration becomes steeper and the reaction attains its maximum value near the particle center. This corresponds to the maximum value of the effectiveness factor. Due to the continuous cell growth, cells which were exposed to optimum substrate concentrations are consequently exposed to concentrations which limit or can not support cell growth. As a result, the intra-particle position where the specific growth rate attains its maximum is shifted progressively away from the particle center towards the particle surface. Apart, therefore, from the shift of the intra-particle location of the optimum cell growth rate towards the surface, the number of cells near the core starts decreasing. As a consequence, the parameter $C_{x,s}/C_{x,ave}$ starts increasing. Eventually, a point is reached where the rate of cell growth near the surface is counter-balanced by the rate of cell decline near the center resulting in an almost constant value for the particle Thiele modulus and a negligible variation in the value of the effectiveness factor.

CONCLUSION

The use of a dynamic diffusion reaction model has demonstrated that the intra-particle position where the specific growth rate attains its maximum, during the course of a substrate inhibited fermentation, is shifted progressively away from the particle center towards the particle surface.

REFERENCES

1. Crick F: Diffusion in Embryogenesis. Nature 1970; 225:420-422.
2. Wada M, Kato J and Chibata I. Continuous production of ethanol using immobilized growing yeast cells. Eur J Appl Microbiol Biotechnol 1980; 10:275-287.
3. Lee TH, Ahn JC and Ryu DDY. Performance of an immobilized yeast reactor system for ethanol production. Enzyme Microb Technol 1983; 5:41-45.
4. Furusaki S, Seki M and Fukumura K. Reaction characteristics of immobilized yeast producing ethanol. Biotechnol Bioeng 1983; 25:2921-2928.
5. Seki M and Furusaki S. Effect of intra-particle diffusion on reaction by immobilized growing yeast. J Chem Eng Japan 1985; 18:461-463.
6. Osuga J, Mori A and Kato J. Acetic acid production by immobilized *Acetobacter aceti* cells entrapped in a κ-carrageenan gel. J Ferment Technol 1984; 62:139-49.
7. Gosmann B and Rehm HJ. Influence of growth behavior and physiology of alginate-entrapped microorganisms on the oxygen consumption. Appl Microbiol Biotechnol 1988; 29:554-559.
8. Kush J and Rehm HJ. Regulation aspects of roquefortine production by free and Ca-alginate immobilized mycelia of *Penicillium roqueforti*. Appl Microbiol Biotechnol 1986; 23:394-399.
9. El-Sayed AHMM and Rehm HJ. Morphology of *Penicillium chrysogenum* strains immobilized in Ca-alginate beads and used in penicillin fermentation. Appl Microbiol Biotechnol 1986; 24:89-94.
10. Mitani Y, Nishizawa Y and Nagai S. Growth characteristics of immobilized yeast cells in continuous ethanol fermentation with forced substrate supply. J Ferment Technol 1984; 62:401-406.
11. Shinmyo A, Kimura H and Okada H. Physiology of α-amylase production by immobilized *Bacillus amyloliquefaciens*. Eur J Appl Microbiol Biotechnol 1982; 14:7-12.
12. Burril HN, Bell LE, Greenfield PF and Do DD. Analysis of distributed growth of *Saccharomyces cerevisiae* cells immobilized in polyacrylamide gel. Appl Environ Microbiol 1983; 46:716-721.
13. Luong JHT. Generalization of Monod kinetics for analysis of growth data with substrate inhibition. Biotechnol Bioeng 1987; XXIX:242-248.
14. Kurosawa H, Matsumura M and Tanaka H. Oxygen diffusion in gel beads containing viable cells. Biotechnol Bioeng 1989; 34:926-932.
15. Chang HN and Moo-Young M. Estimation of oxygen penetration depth in immobilized cells. Appl Microbiol Biotechnol 1988; 2:107-12.
16. Gosmann B and Rehm HJ. Oxygen uptake of microorganisms entrapped in Ca-alginate. Appl Microbiol Biotechnol 1986; 23:163-167.

THE FUTURE

The clearest message to emerge from our studies is that viable cell immobilization may have now reached a stage in its development wherin it can be applied to improve the performance of production-scale industrial bioprocesses, at least for certain biological systems. The technology, an academic curiosity in the late 1970s, is now being turned into a practical fermentation alternative, albeit at a rather slow pace. In our laboratory, we have been rather fortunate to have witnessed the transition of BSP technology to the industrial scale and, much more importantly, the stages associated with this transition. However, industrial applications of the technology seem to be the exception rather than the rule. Why have the deliverables of cell immobilization not been commensurate with the substantial worldwide academic effort that went into and is still going into it?

Over the last two decades, researchers have been rather busy discussing the strengths and opportunities offered by viable cell immobilization and rather reluctant to explore its weaknesses or, indeed, the threats imposed by competing technologies. What people failed to realize is that the advantages offered by cell immobilization are largely of an engineering nature and, as such, cannot possibly compete with the potential for process improvement offered by the underlying biological system. They celebrated the merits of introducing heterogeneities into the bioreactor but seem to have forgotten that gradients of any sort are not particularly welcome by the industrial practitioner who has to control them. Can the complexity introduced to the biological system by immobilization be justified when our knowledge of and our ability to harness the underlying biological system itself is, often, so poor?

Introducing cell immobilization to bioprocesses is more radical a process change than we are often led to believe in the published literature. The risk involved in switching from conventional stirred tank bioreactors to immobilized cell bioreactors is not as low as people often claim and, as such, must command a rather high rate of return from the industrialist. Quite clearly, this will not always be possible. Cell immobilization may offer generic advantages, but the magnitude of the economic benefits of these advantages is very much system-specific. The future lies with only a fraction of these processes, certainly not the ones which ignore the potential of the underlying biological system, but the ones which add to it. Immobilizing genetically-engineered cells is, perhaps, a case in point.

INDEX

Page numbers in italics denote figures (f) or tables (t).

DATE DUE

JUN 2 3 1998		
JUL 1 1 1998		
JAN 0 8 1999		
AUG 0 7 1998		
DEC 1 0 1998	FEB 1 7 1999	
	MAR 2 6 1999	
AUG 2 0 1998		
AUG 3 0 1998		
SEP 2 6 1998		
NOV 0 9 1998		
JAN 2 7 1999		
		Printed in USA

HIGHSMITH #45230